Springer Biographies

More information about this series at http://www.springer.com/series/13617

Jan Guichelaar

Willem de Sitter

Einstein's Friend and Opponent

 Springer

Jan Guichelaar
Amsterdam, The Netherlands

This book draws, in part, on the author's earlier work in Dutch: "De Sitter, Een alternatief voor Einsteins heelalmodel", Veen Magazines, Amsterdam, 2009. [De Sitter, An alternative for Einstein's model of the universe].

ISSN 2365-0613   ISSN 2365-0621   (electronic)
Springer Biographies
ISBN 978-3-030-07486-9   ISBN 978-3-319-98337-0   (eBook)
https://doi.org/10.1007/978-3-319-98337-0

This Springer imprint is published by the registered company Springer Nature Switzerland AG
The registered company address is: Gewerbestrasse 11, 6330 Cham, Switzerland

This book is dedicated to my children Jan and Clasina

Preface

It is a strange chance that, of his most conspicuous contributions, one should relate to the Jovian system—first fruits of the invention of the telescope—and the other to the remotest systems that the telescope has yet revealed.

Arthur S. Eddington[1]

At the wall in the drawing room of my grandfather's house, next to the primary school where he was headmaster, a famous photograph of the Andromeda Nebula hung in the 1950s. As the son of a barber in a small Frisian village, the only way for the bright young boy to study was to attend a teacher training college, in The Netherlands sometimes called *the university for the poor*. His interests and capacities went further than the primary school teachers' level. In his youth, the universe consisted only of the visible galaxy, in the years that the Milky Way was still visible nearly everywhere in the Netherlands. He explained the Milky Way to me by roughly drawing evenly distributed dots, the *stars*, on two glass saucers. Putting one upside down onto the other, he had made a model of our galaxy. If you looked perpendicular on the saucers, you saw the stars at their normal distances. But if you looked in the plane of the two saucers, you saw the dots seemingly close to one another, forming a *Milky Way*. It was my first insight into cosmology.

In his youth, the complete universe consisted of our own galaxy, nobody supposed anything outside it. But in the 1920s, the first distances to the nebulae were measured and they turned out to be complete galaxies of their own, at extremely large distances. It must have made a great impression on the young teacher. Roughly thirty years later, his interests in astronomy had not abated. One of his daughters presented him with a telescope, which had a maximum magnification of 150. As a delighted young boy he, an older man not long before his retirement, put up his telescope night after night to see the wonders Galileo had seen 350 years

---

[1] Arthur S. Eddington in his obituary on Willem de Sitter in *Nature* (CXXXIV, December 15, 1934).

The Andromeda Nebula

earlier: the mountains on the Moon, the phases of Venus and the four great moons of Jupiter. Some of these sights I remember when I stayed there during holidays, at the teacher's house in a small village in the eastern part of the province of Groningen, in those days not yet connected to the main water supply, nor the gas and electricity networks. Shortly before his death, he presented the telescope to me. As a theoretical physicist, I was not active as an amateur astronomer and the box with the telescope has been stored in the garage or cellar most of the time. The last time I used the telescope was in June 2017, when friends of mine had taken the telescope to their second home in Liguria in Italy, ten miles north of Ventimiglia. When I visited them a couple of days, we put up the telescope, and on a perfectly clear night, we had a beautiful view of Jupiter and his four Galilean moons, two on either side of the small disc. I dare say that my grandfather played a role in arousing my interest in astronomy.

So it is not surprising that I chose as my topic for a biographical research the life and work of the Dutch astronomer Willem de Sitter (1872–1934), because his work knew two main topics: a new theory for the four Galilean moons of Jupiter on the basis of classical celestial mechanics, and the description of the later called De Sitter Universe on the basis of Einstein's general theory of relativity.

Explanatory Notes

- The first version of this book was written in the years before 2009. In that year, a shortened version of the current book was published in Dutch: *De Sitter, An alternative for Einstein's model of the universe* (Dutch: *De Sitter, Een alternatief voor Einsteins heelalmodel*); Veen Magazines, Amsterdam, 2009. The present biography is an extended version, including a number of new aspects. The complete scientific account with all necessary footnotes and literature references, for which there was no place in the Dutch book, is incorporated in the present book.
- In the ongoing texts, the references to letters (mostly from and to De Sitter) and a number of (dated) documents are given between square brackets. If the writer and receiver of the letter(s) are clear from the text, only the abbreviation of the archive—and sometimes a number and code—and the date of the letter are given. If the name is not clear, it is given also between the square brackets.
- A number of drawings were made using the drawing program GeoGebra.
- The photographs are for a large part, with the permission of the Leiden Observatory, taken from the archive of De Sitter in the University Library of Leiden. Another part is taken from several family archives with the permission of the owners. A number of pictures and photographs are from the author's personal archive. There are a number of photographs from other sources, and a few photographs (all more than a century old) were free of rights available in the public domain. The sources are given in the text as completely as possible.

Amsterdam, The Netherlands   Jan Guichelaar
2018

Acknowledgements

In the first place, I am indebted to Anne J. Kox, emeritus Pieter Zeeman professor in the history of physics at the University of Amsterdam. In 2006, he mentioned to me the lacuna in the research into the work of Willem de Sitter and gave me advice during my research. For many stimulating discussions, I owe many thanks to Bastiaan Willink, retired sociologist and historian of science, and a close friend since 1951.

Many thanks I owe to a number of family members of Willem de Sitter, who provided me with information. First, W. Reinold de Sitter (who died in 2009), son of Willem de Sitter's son and geologist Lamoraal Ulbo, who told me in a number of conversations frankly and with humour a lot of relevant facts concerning the De Sitter family. Besides, I could use the result of his own work on his grandfather.[2] I am also indebted to Reinold's brother Lamoraal Ulbo and sister Tjada van den Eelaart-de Sitter, to Willem Jan de Sitter, son of Willem de Sitter's son and astronomer Aernout, and to Ernst de Sitter, grandson of Willem de Sitter's brother Ernst Karel Johan. I also wish to express my thanks to other members of the family, who provided me indirectly with useful information.

The original De Sitter Archive was stored in tens of boxes in the cellars of Leiden Observatory, waiting for over seventy years for an inventory to be made. A few times parts of this work were undertaken. De Sitter's correspondence with Einstein was traced and published. In 2008–2009, historian of science David Baneke made the complete inventory and now the archives are kept in the Leiden University Library.[3] I am indebted to Baneke for his support in finding relevant data in these archives. As a general reference to Dutch astronomy in the twentieth

---

[2]Reinold de Sitter wrote *Grandfather, a charcoal sketch* (privately published, present in the De Sitter Archive at the Library of Leiden University).

[3]The De Sitter archives are, also digitalized, present in the Leiden University Library, Special Collections, as *Leiden Observatory Archives, directorate W. de Sitter* (Collection guide written by David Baneke, 2010).

century, Baneke's book *De Ontdekkers van de Hemel* (*The Discoverers of the Heavens*) gives ample information.[4]

Adriaan Blaauw (1914–2010, till his death the nestor of Dutch astronomy) provided me in a number of conversations with information, in particular concerning life at Leiden Observatory in the thirties of the twentieth century. Blaauw started his studies in astronomy after a conversation in 1932 with De Sitter, at the end of which De Sitter nodded positively to the young Blaauw, who decided then to choose for Leiden and not Amsterdam. Blaauw is the only person I know who could recollect a conversation with De Sitter (*with his sloppy jacquet over his thin shoulders*).[5] In retrospect, I owe a lot to Blaauw.

For the analysis of De Sitter's contribution to the relativity theory, the Collected Papers of Albert Einstein,[6] in particular the work of Michel Janssen[7] therein, and the thesis of Stefan Röhle[8] have been of great importance for me. Röhle's dissertation is a gold mine of personal details, relevant archives and a nearly complete bibliography.

I owe many thanks to a large number of archival workers of Dutch archives:

- De Sitter Archives in Leiden Observatory (later Leiden University Library);
- North Holland Archives[9] (Haarlem);
- University Libraries of Groningen and Amsterdam;
- Archives of the province of Gelderland[10] (Arnhem);
- Regional Archives[11] (Leiden);
- Museum Boerhaave (Leiden);

and of a number of archives outside The Netherlands:

- University Library of Ghent (Belgium);
- Union Internationale Astronomique (International Astronomical Union) in the Institut de France in Paris (France);
- Royal Geographical Society in London (England);
- University Library of Cambridge (England);

---

[4]Baneke, D., *De Ontdekkers van de Hemel, De Nederlandse Sterrenkunde in de Twintigste Eeuw* (*The Discoverers of the Heavens, The Dutch Astronomy in the Twentieth Century*); Prometheus, Bert Bakker, Amsterdam, 2015.

[5]There are a few moving images of De Sitter, smoking a pipe and reading his post. They are part of an amateur film made in 1933 on the occasion of 300 years Leiden Observatory, now for restoration in the Netherlands Institute for sound and vision in Hilversum (The Netherlands).

[6]In an ongoing research process, the *Collected Papers of Albert Einstein* are being published by Princeton University Press.

[7]The results of Janssen's work on the controversies between Einstein and De Sitter concerning a number of relevant points in Einstein's cosmological ideas are included in Michel Janssen and Christoph Lehner (editors), *The Cambridge Companion to Einstein*, "No success like failure …": Einstein's Quest for General Relativity; Cambridge University Press, 2014.

[8]Stefan Röhle wrote his thesis at the Johannes Gutenberg University in Mainz in 2007: *Willem de Sitter in Leiden—Ein Kapitel in der Rezeptionsgeschichte der Relativitätstheorien*.

[9]Dutch: Noord Hollands Archief.

[10]Dutch: Gelders Archief.

[11]Dutch: Regionaal Archief.

- Huntington Library in Pasadena, Los Angeles (USA);
- Institut for Videnskabsstudier of the University of Århus (Denmark);
- Durham University in Durham (England);
- Council for Scientific and Industrial Research (CSIR) in Pretoria (South Africa).

For information, cooperation and suggestions (in conversations, by telephone, by letter, by e-mail) in support of this research I thank:

- Bertha Clemens Schröner (who died ca. 2010, Oegstgeest, the Netherlands), who for a period lived as a guest in the director's large living quarters in the old Leiden Observatory (then inhabited by Jan Oort, a successor of De Sitter);
- Dirk van Delft, Director of Museum Boerhaave in Leiden;
- Ed van den Heuvel, Professor and Former Director of the Astronomical Institute Anton Pannekoek in Amsterdam;
- Vincent Icke, Professor of theoretical astronomy at Leiden Observatory;
- Annette Joubert of the CSIR in Pretoria (South Africa);
- Jet K. Katgert, Leiden Observatory;
- Helge Kragh, Emeritus Professor of the history of science of Århus University (Denmark);
- Johan van Kuilenburg, Dutch astronomer;
- H. van Loo, Judge at the court of justice in Arnhem;
- Jan Lub, Associate Professor of astronomy at Leiden Observatory;
- Frans van Lunteren, Professor of history of science;
- Geart van der Meer, Frisian poet and translator;
- N. Nelissen, Teacher at the City Grammar School[12] in Arnhem;
- Matteo Realdi, historian of science;
- W. Suermondt, relative of De Sitter's wife, Rotterdam;
- Lambert Swaans, Old Leiden Observatory;
- Brian Warner, Professor at the Department of Astronomy, University of Cape Town (South Africa);
- Henk Venema (who died in 2011) and René Luijpen for carefully reading and improving (parts of) my original Dutch texts;
- In particular, I thank my wife Ria Koene and daughter Clasina Guichelaar, both teachers of English, for their meticulous text corrections, and my son-in-law Henk Schuitemaker for adjusting and improving a number of the photographs.

---

[12]Dutch: Stedelijk Gymnasium.

Contents

Abbreviations

AG	Astronomische Gesellschaft
AH	Archives of Hertzsprung, History of Science Archives, Center for Science Studies, Aarhus University, Denmark
AIP	Archives of the IAU in the Institut d'Astrophysique de Paris
AL	Archives of Lorentz in the North Holland Archives in Haarlem
AN	Astronomische Nachrichten
AOL	Annals of the Observatory in Leiden (Dutch: Annalen van de Sterrewacht te Leiden)
AP	Annalen der Physik
ASL	Leiden Observatory Archives, directorate W. de Sitter, in the University Library Leiden
AUL	Archives of Leiden University, in the University Library Leiden
BA	British Association for the Advancement of Science
BAN	Bulletin of the Astronomical Institutes of the Netherlands
BG	Board of Governors
CPAE	*Collected Papers of Albert Einstein*, Princeton University Press, since 1987
CPD	Cape Photographic Durchmusterung
DUL	Archives of Durham University Library
hbs	Higher Civilian School (Dutch: Hogere Burgerschool)
HMW	Archives of the Holland Society of Sciences, in the North Holland Archives in Haarlem. (Dutch: Hollandsche Maatschappij der Wetenschappen)
IAU	International Astronomical Union
IRC	International Research Council
IUGG	International Union of Geodesy and Geophysics
MNRAS	Monthly Notices of the Royal Astronomical Society
NAC	Dutch Astronomers Society

NAW1	Archives of the (Royal) Netherlands Academy of Sciences, in the North Holland Archives in Haarlem. (Dutch: (Koninklijke) Nederlandse Akademie van Wetenschappen)
NAW2	Reports and Proceedings of the (Royal) Netherlands Academy of Sciences, in the International Institute of Social History in Amsterdam
PZ	Physikalische Zeitschrift
RAS	Royal Astronomical Society
RGO	Archives of the Royal Greenwich Observatory in the University Library in Cambridge
RGS	Archives of the Royal Geographical Society in London

Chapter 1
Astronomy in Leiden till De Sitter's Arrival

> *In those days the here performed observations enjoyed, under*
> *Kaiser and Bakhuyzen, the well-earned fame of being among*
> *the most accurate of all observatories in the world.*
> Willem de Sitter in a speech on the occasion of the inauguration
> of the reorganized Observatory in Leiden by his excellency the
> Minister of Education, Arts and Sciences on 18 September 1924

The First Two Centuries

De Sitter called Jacob Gool in his brochure of 1933, on the occasion of 300 years of Leiden Observatory,[1] the founding father of the Observatory (De Sitter 1933). Jacob Gool, or latinized Golius (1596–1667) studied mathematics in Leiden with Willebrord Snel van Royen, latinized Snellius (1580–1626), who was professor of mathematics at Leiden University and famous for his law on the refraction of light rays. In order to study the works of the Greek mathematician Apollonius in Arabic translations, Gool studied Arabic and became a prominent Arabist as well. In the morning he lectured on mathematics, physics or astronomy and in the afternoon on Arabic language or literature. Descartes became one of his students in mathematics. In 1632 Gool persuaded the University of Leiden to buy the so-called Quadrant of Willem Blaeu (1571–1638; cartographer) from Snellius and to pay for a small wooden structure on the roof of the Academy building to be used as a modest Observatory.

Gool did measurements on comets, eclipses and planets. With the help of a quadrant the angular distance of a celestial body above the horizon or from a certain other direction could be measured. The direction of the celestial body was established by means of a narrow pipe. The angles could be read on a graduated arc.

---

[1]Dutch: Sterrewacht.

© Springer Nature Switzerland AG 2018
J. Guichelaar, *Willem de Sitter*, Springer Biographies,
https://doi.org/10.1007/978-3-319-98337-0_1

Fig. 1.1 Blaeu's quadrant in Rijksmuseum Boerhaave in Leiden. Photograph taken by the Author

After a renovation and enlargement at the end of the seventeenth century the Observatory stayed unaltered till the beginning of the nineteenth century. By several successive professors of astronomy new instruments were regularly bought. Willem Jacob 's Gravesande (1688–1742) for instance bought a reflecting telescope in 1736, made by one of the first commercial telescope makers George Hearne of London (Zuidervaart 2007). 's Gravesande studied law in Leiden and became a barrister in The Hague. Besides he was a mathematician, physicist and astronomer. He became professor in Leiden in physics and astronomy in 1717, the first professor with an explicit assignment to teach astronomy. He visited Newton in England and propagated his work in The Netherlands. But in those years there was no systematic astronomical research yet. The Observatory was mainly used for educational aims and to impress interested citizens. Plans to build a larger and better Observatory kept on being rejected.

Frederik Kaiser

Not until the nineteenth century a revolutionary change was brought about by the astronomical talent of Frederik Kaiser. Kaiser's father had come from Germany to The Netherlands, where Frederik was born in Amsterdam in 1808. After his father's demise when Frederik was eight his uncle Johan Frederik Keyser took on the care of the young Frederik. The uncle was an enthusiastic amateur astronomer and it was from his uncle that the young Frederik learned the principles of astronomy and the

Fig. 1.2 Jacobus Golius, founding father of Leiden Observatory, professor at Leiden University in mathematics and eastern languages. Photograph from the Leiden Observatory Archive

relevant mathematics. When Frederik was only fifteen also his uncle died. The astronomer Gerrit Moll (1785–1838) from the Observatory of Utrecht, who himself had also been taught the principles of astronomy by Frederik's uncle, found the young Kaiser a job in 1826 as an observer at the Leiden Observatory with earnings of 800 Dutch guilders yearly. His appointment had been made without the

Gezigt op de ACADEMIE, te Leyden.

Fig. 1.3 Academy building in Leiden with observatory platform. Photograph from the Leiden Observatory Archive

knowledge of the just appointed professor of astronomy Pieter Uylenbroek (1772–1844). This was one of the reasons of their mutual poor relationship.

Kaiser was ambitious, but he lacked a thorough academic education. In his studies he did only reach the so-called first exam (cf. bachelor degree). But in 1835 at last he received recognition as a result of his excellent work on the comet of Halley. He had done his observations from the attic of his house, where he had removed some tiles. His prediction of the orbit and perihelium passing of the comet was much more accurate than those of foreign astronomers. For those results he received a doctorate *honoris causa*. That paved the way for a further career (Zuidervaart 1999). In 1837 he was appointed as lecturer and director of the Observatory. In 1840 his appointment followed as extraordinary professor and in 1845 as ordinary professor. In 1839 he complained heavily about his salary not being raised and he blamed Uylenbroek. He was notoriously complaining about nearly everything. He drastically improved the old Observatory on the roof of the Academy building, but the money for building a new 'proper' Observatory was still lacking. He started a programme of regular observations on double stars, asteroids, and comets. He popularized astronomy in large parts of the population by popular

Fig. 1.4 Frederik Kaiser, father of modern Dutch astronomy and builder of the new Observatory. Photograph from the Leiden Observatory Archive

writings and lectures. In particular his popular book *The Starry Heavens*[2] became a big success, although there is not a single picture in the book. After the successful first edition a second followed soon in 1850. But the developments in the astronomy in those years went so fast that the publisher urged Kaiser to write a completely revised third edition by 1860. He had to do this in between many pressing tasks, of which the preparations for the building of a new Observatory certainly were the most demanding. He complained heavily about this lack of time:

> I was forced to lean on my memory and I have had to experience, how much the mind is numbed by the continuous tribulations of life (Kaiser 1860).

But complaining was his second nature. The book became extremely popular and Kaiser received a lot of fan mail. From Amsterdam he received a poem, in which he was compared to Maarten Luther, who had translated the Holy Scriptures into German for the masses. From this poem (in translation):

> But with a clear voice a well of knowledge
>
> rises in the people's face, in which flow it's baptized.
>
> And with the eye directed to the visible infinity
>
> you open, like Luther did the Word, a different Bible.

---

[2]Dutch: *De Sterrenhemel*. See Kaiser (1860).

In the meantime, in spite of complete silence to all requests from the side of the government, Kaiser's popularity made it possible that a committee of private persons raised 26,000 Dutch guilders for the building of a new Observatory. By far not enough: the estimate was 112,500 guilders. But at last the train was moving. Even the students raised 800 guilders. It may be clear that Kaiser aimed at a substantial project, in view of the fact that the building of the modest observatory in Utrecht in 1854 had cost 12.583 guilders, one tenth of the estimate for Leiden (De Jager et al. 1993). But it was the appointment of Gerrit Simons to Minister of Internal Affairs (in whose portfolio Education fell) that led to real progress. Simons had been an observer with Moll in Utrecht from 1825 till 1831, where he had studied mathematics and physics (De Jager et al. 1993). Of course he was well disposed towards astronomy and he included the money for the building of the Observatory in his budget for 1857. But his budget was voted down and not much later Simons resigned. His successor ridder (*knight*) Van Rappard was an admirer of Kaiser and had been a member of the fund raising committee. The money was still included in the budget for 1858. In his function as Rector[3] of the University Kaiser used his influence to choose the southern part of the *Hortus Botanicus* (botanical garden) as location for the new Observatory. In his eyes that garden was in a poor condition. That was completely against the wish of professor of botany W. F. R. Suringar (1832–1898), director of the botanical garden. It became an indecorous struggle. In the Student's Almanac of 1857 a cartoon was shown with Kaiser trying to pull the key of the botanical garden out of the hands of the struggling professor Suringar, with a botanical box hanging round his neck (Van Herk et al. 1983). Kaiser won the battle. The Observatory in Pulkovo near St. Petersburg provided a model for the design of the Belgian architect Henri Camp. Camp was also the architect of the physics laboratory in Leiden (now the seat of Leiden Law School), later named after Nobel prize winner H. Kamerlingh Onnes (1853–1926). But Camp did not know much about astronomical requirements and to the great annoyance of Kaiser he went too much his own way. The Observatory was built for 132,000 guilders. For the stability of the new building more than 1500 wooden piles were used, of which more than 100 to provide a vibration-free support for the meridian circle (the most important telescope). In 1860 the new Observatory was opened. Of course Kaiser complained unremittingly with everyone who wanted to listen, and in pathetic language as De Sitter would write later, that the whole building process went too slowly and that he received too little money for staff and research (De Sitter 1923).

The main theme of research Kaiser chose was the fundamental astronomy, so mainly the accurate determination of the location of stars. This line of research had been decided for long before the architectural drawings were made. Fundamental determination of star locations was a field of research with which a smaller Observatory in a small country could also advance the astronomical science. But given that, perfection had to be achieved. The main instrument to reach that was the meridian circle made by Pistor and Martins in Berlin and taken into use in 1861.

---

[3]In Dutch the term *Rector Magnificus* is used.

Meridian Circle

A meridian circle is a telescope that can only turn in the meridian plane (the plane through the poles and the local zenith; thus north-south). It was primarily used to measure exact time, when a star passed the local meridian. Moreover, the direction of the telescope yielded (with the help of microscopes on the circles attached to the telescope) the latitude of the star (in degrees °, minutes ' and seconds " of arc).

In order to measure the time the greatest accuracy was aimed for. The Hohwü 17 clock of 1860 was, as De Sitter later remarked *exceptionally good, the principle clock of the Observatory* (De Sitter 1933). The clock ran for more than forty years with only a single cleaning. Measuring the time was done electrically with a tapping-key from 1866 onwards. The key had to be tapped at the moment the star passed the crosshair in the field of vision of the meridian circle. This method was an enormous improvement in comparison to the old method, where the point of time of the passage was estimated while listening to the ticking of a clock (the eye and ear method). For Kaiser precision measurements were his first priority. In his eye they were more important than making new discoveries. Better more accurate measurements of the place and orbit of a known planet than discovering a new one. With this attitude Kaiser seems to have influenced many researchers after him: his successors Hendrik and Ernst van de Sande Bakhuyzen and perhaps also the low-temperature physicist Heike Kamerlingh Onnes (1853–1926), who had chosen *Through measuring to knowledge* as his research motto. The two revolutionary new methods of measurement in astronomy developing in those years, photography and spectroscopy, were still in their infancy. For Kaiser that was not enough for a place in his research programme. A broader programme would also have required bigger and much more expensive instruments.

Kaiser transformed the Dutch astronomy, coming from an amateur status, into a professional science, with the marks of a complete institute, the accompanying instruments, staff and a scientific research programme. He became a great scientist, greatly admired by a lot of people. In 1871 he fell ill. He wrote in desperation, slightly confused but typically Kaiser, to bookseller Sulpke:

> The greatest idiocy that ever could have entered my brain was my wish to found a seat of astronomy in Leiden.[4]

He did not recover and died in 1872. Of course a poem was written in his memory, which reads in translation:

---

[4]Archive of MEOB, Marine Electronic and Optical Company (Dutch: Marine Elektronisch en Optisch Bedrijf); Dutch Institute for Military History (Dutch: Nederlands Instituut voor Militaire Historie), The Hague.

While all men slumber, being awake and on his post,

Observing the horizon in the solemn nightly hour,

Enclosing the shining cupola of Gods immeasurable skies,

With all the teeming armies of the suns,

So, KAISER, did we know you.

He could have lived with it. Due to Kaiser the astronomy in Leiden counted internationally: in 1875, three years after Kaiser's death, the meeting of the important Astronomische Gesellschaft (Astronomical Society) of Germany was held in Leiden.

A Few Observers

A number of observers at Leiden Observatory has had a clear influence on the professional development of the other observatories in The Netherlands.

Utrecht: Jean Abraham Chrétien Oudemans (1827–1906)

As a young student Oudemans attended Kaiser's lectures and took his doctoral degree in 1853 under his guidance. After that he became an observer at Leiden Observatory. In 1856, still before the building of the new Observatory, at the recommendation of Kaiser he left for Utrecht as an extraordinary professor in astronomy. The post did not suit him and then he worked for years in the Dutch East Indies, inter alia on the triangulation of the island of Java. In a triangulation, with the help of triangular measurements and astronomical sightings, a country is mapped exactly. At the succession of Kaiser in 1872 he was passed by Hendrik van de Sande Bakhuyzen. The University of Utrecht asked him to return to his old chair. Thereupon he was director of Utrecht Observatory from 1875 till 1898. He taught a lot and revised Kaiser's *De Sterrenhemel*. In 1874 Oudemans led an expedition to the island of Réunion in the Indian Ocean in order to observe the transit of Venus in front of the sun. The expedition became a failure by too many clouds. During a transit the shadow of Venus is visible on the sun's disc. With these observations the distance from the earth to the sun could be measured. Members of the expedition were also Kaiser's son Pieter and Van de Sande Bakhuyzen's younger brother Ernst. In 1898 Oudemans was succeeded in Utrecht bij Albertus A. Nijland (1868–1936), who stayed director until his death and even played an interesting role in the process of De Sitter's appointment as director of Leiden Observatory in 1918 (De Jager et al. 1993; Pyenson 1989).

> **Venus Transit**
> From good observations of a Venus transit (Venus passing between the earth and the sun, creating a dark spot on the sun's disc) the distance from the earth to the sun can be calculated. This procedure was first suggested by Edmund Halley at the Venus transits of 1761 and 1769. The distance of the earth to the

sun was necessary to determine the size of the planetary system. The failure of the expedition in 1874 did not result in a Dutch expedition in 1882. A number of other countries did organize an expedition. Only in 2004 and 2012 there were the next two transits. The transits occur in couples with a time distance of 8 years. And between the couples a period of more than a century.

Groningen: Jacobus Cornelius Kapteyn (1851–1922)

Kapteyn, the later teacher of De Sitter, became an observer in Leiden after his doctorate in 1875, under Kaiser's successor Hendrik van de Sande Bakhuyzen. That was his period of apprenticeship as an experimental astronomer. As a result of the new law of 1876 on higher education in The Netherlands the University of Groningen received a chair for astronomy. Kapteyn was appointed there in 1878 and under his guidance astronomy blossomed in Groningen. At the end of the century the Kapteyn's Sterrenkundig Laboratorium (Astronomical Laboratory) had surpassed Leiden in importance. Leiden had slowly fallen into a decline by constantly holding on to sheer location measurements of celestial bodies. For a short biography of Kapteyn see Chap. 6.

Amsterdam: Antonie Pannekoek (1873–1960)

Pannekoek started his studies in mathematics and physics and chose for astronomy after his bachelor's exam. He passed his master's exam in 1895. In 1898 he became an observer at Leiden Observatory, where he wrote his doctoral dissertation with Hendrik van de Sande Bakhuyzen as supervisor: *Untersuchungen über den Lichtwechsel Algols* (*Researches on the changing luminosity of Algol*), then a very modern subject. In those days he was already an outspoken socialist. One of the short propositions in his dissertation[5] was:

> The progress of the natural sciences is hampered by the continuation of the capitalist production.

In 1906 Pannekoek left for Germany. According to Pannekoek at the Observatory there was always executed a next series of meridian measurements without making the necessary reductions. The new astronomy (photography and spectral analysis) still stayed outside the scope of the research programmes in Leiden. In Pannekoek's book *Reminiscences* (Pannekoek et al. 1982) he made two striking remarks. About the period after the appointment of Ernst van de Sande Bakhuyzen as his brother's successor:

---

[5]Every dissertation in The Netherlands has a number of short propositions or comments on a variety of scientific or social subjects at the end.

Then everything went even slacker.

And about Willem de Sitter's decision after his appointment not to make reductions of all old measurements:

So, all that work for which I worked so hard, swept under the carpet; and right so.

After his return to The Netherlands Pannekoek became, with the support of De Sitter, a private lecturer in Leiden. When De Sitter asked him, after his appointment as director of Leiden Observatory in 1918, to become one of the deputy directors, his appointment was blocked by the government, as a result of his explicit socialist ideas. In 1919 he was then appointed as a lecturer at the University of Amsterdam, under the jurisdiction of the city council of Amsterdam and not of the government. In 1925 he became an extraordinary professor and in 1932 a full professor. In 1925 De Sitter and Ejnar Hertzsprung, deputy director in Leiden, then proposed Pannekoek successfully as a member of the Dutch Academy of Sciences [ASL, from Hertzsprung, 1-4-1925]. Also after 1919, already at work in Amsterdam, Pannekoek kept being in close contact with De Sitter. He borrowed a lot of things from Leiden in his first years in Amsterdam: a Repsold measuring device, books, magazines. Moreover he often asked De Sitter's advice concerning his current investigations. At first he worked on stellar astronomy and the structure of the universe, in line with Kapteyn's research, later he focussed on the evolution of stars. In 1921 he founded The Astronomical Institute of the University of Amsterdam (now called the Anton Pannekoek Institute for Astronomy). On 4 October 1918 he gave his public lecture, titled *The Evolution of the Universe* (Pannekoek 1918). In this lecture he stated, on the basis of the conservation laws of mass and energy: the world did not originate and will not perish, there is only evolution. He was occupied with the universe, although he neither mentioned the general theory of relativity, published a few years earlier, nor the first articles about cosmology of Albert Einstein and De Sitter. Probably those were too recent (Pannekoek et al. 1982).

Kaiser as Auditor of the Empire's Nautical Instruments

In addition to his astronomical activities Frederik Kaiser was appointed in 1858 as *Auditor of the Empire's Nautical Instruments*, a title he coined himself. It is not remarkable that the care for accurate nautical instruments was entrusted to an astronomer. In order to determine a ship's location at sea the precise place of a number of celestial bodies and the instruments to measure these played an important role. With this job Kaiser earned an additional 1000 Dutch guilders a year, and he equipped a small room in the Academy building for the purpose. It was not long before he started to complain about the workload, he felt a warehouse manager. In 1860 he managed to appoint an assistant auditor, a clerk and a servant. Not completely free of nepotism he managed to get his son appointed to assistant auditor. Kaiser advised the Secretary of the Navy regarding purchase and inspection of all kinds of instruments. From the attic of the Academy building at Rapenburg canal in Leiden the Auditing Office moved to the new Observatory, where in 1873 annexes for the Office were built on both sides of the main entrance. The instruments about which father and son advised were diverse: chronometers, devices for measuring distances, clinometers (to measure inclination angles), magnetogalvanic torpedo detectors, compasses, sextants (for measuring the angle between a celestial body and the horizon), prism circles (angle measurements) and mercury horizons (artificial horizons). An extremely important action was giving exact time signals. From 6 September 1858 twice a week, on Tuesday and Friday nights, a signal was sent through the telegraph cable net, by which network a number of harbours had to be connected, for the benefit of the navy and the merchant navy. In the new Observatory the signal was taken from the astronomical pendulum clock of the Amsterdam clockmaker A. Hohwü, the nr. 17 from 1860. This clock, at its turn, was synchronized with the help of the meridian circle. The signals were nearly always correctly transported. Only once no signal came through. A few anglers had dredged up the cable in the city's outer canal near the Observatory and had taken home part of it. The sending of telegraphic time signals continued into the directorate of Willem de Sitter and stopped in 1921. After that radiographic signals came from Paris, which could be received by anybody. After Kaiser's death in 1872 he was succeeded as Auditor by his son. The Auditing Office grew and lacked space. It was also an odd man out in the Observatory and interfered with the regular work there. Therefore the Office left and found a new place at the Old Pig Market[6] in Leiden (Spek 1979).

Latitude and Longitude at Sea

One of the biggest problems of seagoing ships in earlier days was the determination of a ship's place at sea, without land in sight. The latitude was measurable well at night: it is equal to 90° minus the angle between the directions of the Pole Star and the zenith (the direction through the centre of the earth and the ship upwards).

But measuring the longitude was much more difficult. That required a good clock, which was calibrated in Greenwich (on the prime meridian, with longitude 0°), indicating 12:00 h sharp at the moment the sun reached its highest point in the meridian. When for instance this clock somewhere at sea indicated 15:00 h at the moment the sun reached its highest point, the earth had rotated $(3/24) \times 360°$ to the east. The longitude of the ship was then 45° western. But good clocks were hard to make. Christiaan Huygens's (1629–1695) pendulum clock did not properly work on a rough sea. The first good operating chronometer clock was designed and built by the Englishman John Harrison (1693–1776). His clock from 1764 was only 1/10 of a second a day slow on a long sea trip (Sobel 2008).

---

[6]Dutch: Oude Varkenmarkt.

Photography

Kaiser's son Pieter Jan later came to work at the Observatory. He started as an amateur photographer. He became an astronomer and took his doctoral degree in 1862 with his father as supervisor. For his thesis, titled *The application of photography to astronomy* (Kaiser 1862), he could not get any photographs from abroad. It was still a time in which photography had hardly entered public and scientific life. In spite of the title there is only one photograph to be found in the thesis: one of the moon, made at the Leiden Observatory in 1862. The use of photography for scientific research in the early days of photography gave considerable difficulties. The long exposure times required the telescope to rotate in co-ordination with the rotation of the earth. However, the engines to realize this had not yet the necessary steady movement. It took up to 1880 before photography was applied on a large scale. In 1862 the young Kaiser did probably not yet have much confidence in it, in view of two of the short propositions at the end of his dissertation:

> Photography has as yet not been able to give essential contributions to astronomy.

> The effectuation of measurements on photographic pictures has to be rejected.

It is not improbable that his father, who vowed to his classical precision measurements, was partly responsible for these short propositions (Rooseboom 1994).

Hendricus (Hendrik) Gerardus van de Sande Bakhuyzen (1838–1923)

Van de Sande Bakhuyzen first took up his studies in Delft, where there was a Royal Academy for Engineers (from 1864 the Polytechnic School, now one of the Technical Universities in The Netherlands). He continued his studies in Leiden and wrote his dissertation in 1863 with Kaiser as supervisor: *On the influence of bending on the heights of celestial lights, measured with the meridian circle*. It dealt with causes of possible measurement errors (Van de Sande Bakhuyzen 1863). To that end the influence of the refraction of light rays in the room and inside the telescope had to be taken into account as well. Striking is his first short proposition:

> In astronomy there does not exist a more fatal error than multiplying the number of instruments, without enough extra hands to operate them.

Van de Sande Bakhuyzen was an old school physicist considering the first words in his dissertation:

> Generally one assumes that material bodies are built of a large amount of small particles (atoms),

and his short propositions about the absolute necessity of the ether and the fact that the existence of elements was unprovable. The dissertation did not give a complete solution of the problem, that would still remain wishful thinking for a long period. Van de Sande Bakhuyzen became a teacher and in 1867 a professor in Delft. He succeeded Kaiser after his death in 1872. The new astronomy, photography and

Fig. 1.5 Hendrik van de Sande Bakhuyzen at the meridian telescope of Leiden Observatory. Photograph from the Leiden Observatory Archive

spectroscopy, had already yielded a number of results in astronomy, but Van de Sande Bakhuyzen was not convinced that these would continue. He was neither convinced that he would get the necessary funding. He chose certainty and continuation of the fundamental astronomy, as initiated by Kaiser. Under Van de Sande Bakhuyzen's regime Leiden became one of the prominent observatories for exact determination of locations, although the reductions of the long measuring programmes not seldom took too much time.

The measuring programmes covered from 1870 till 1876 the stars in a number of A.G.-zones (the German Astronomische Gesellschaft), published by observer Jan Hendrik Wilterdink (1856–1931) in 1902. For a short period Wilterdink was acting director of Leiden Observatory after the death of Ernst van de Sande Bakhuyzen and before the appointment of De Sitter in 1918. From 1879 till 1898 measurements were made of stars near the zenith, in Leiden as well as by the Observatory of Cape Town in South Africa. Identical stars were measured by both observatories. Comparison yielded corrections as a result of the refraction of light rays in the atmosphere. To this aim Van de Sande Bakhuyzen investigated also with the help of balloons the temperature distribution in the atmosphere, which influences the refraction. In fact he investigated the effect of refraction on the measurements of declinations, which also occupied Kapteyn for decades, and even De Sitter with his Kenia expedition in 1931. Declinations of stars measured at different observatories often differ 1 s of arc. It is significant that the reduction of these measurements were only published in 1928 in an article by Jan Oort (1900–1992), nearly half a century after the first measurements. He again ascertained the absolute necessity to measure fundamental declinations with the help of different methods (Oort 1928). The Observatory and the set of instruments were enlarged during Van de Sande Bakhuyzen's regime. The Dutch architect Pierre Cuypers, famous for Amsterdam Central Station and the Rijksmuseum, designed the tower with an observation dome built in 1878 on the west side of the Observatory. The heliometer, which according to the plans would be placed in it, was never realized. Neither were the signs of the zodiac in the design realized on the outside.

Declination and Right Ascension

The place of a star S on the celestial sphere may be given in the equatorial co-ordinate system by the two co-ordinates declination δ and right ascension α. As base we take the earth's equatorial plane.

In the left figure the plane through the north N, a star S and the centre of the earth M is drawn. This plane intersects the equatorial plane along MS'. The declination of the star S is defined as the angle δ = SMS'.

In the right figure the equatorial plane is drawn. In that plane we have the fixed vernal point Y (that is the point where the sun passes in his orbit in the ecliptic the equatorial plane around 20 March). The right ascension α of the star is defined as the angle α = YMS'.

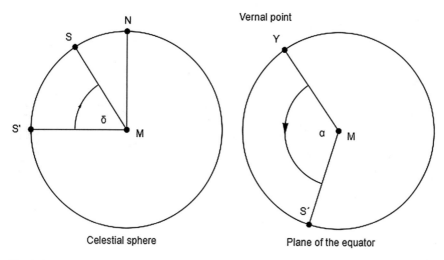

Fig. 1.6 Declination and right ascension

In his researches Van de Sande Bakhuyzen always gave much attention to the greatest accuracy and to (the prevention of) all kinds of systematic (instrumental) and personal measuring errors. In order to start doing measurements an observer had to let the telescope acclimatize to the warmth of the human body for an hour. Van de Sande Bakhuyzen dedicated himself also to the in those years growing international cooperation. He took part in the Paris conference in 1887 where plans were made for the Carte du Ciel (Sky Map), a star catalogue for the complete sky, made with the help of photographs. Van de Sande Bakhuyzen's method of measuring the photographic plates was chosen ignoring those of Gill and Kapteyn. The influence of the old master was great. On the east side of the main building of the Observatory a large photographic telescope was placed in a new tower with dome, also intended to measure star parallaxes.

Parallax
The parallax of a star is used to measure the distance from the sun to that star. Consider a near star N. Seen from the earth with a time lapse of 6 months we can measure the angle E_1NE_2. This angle is by definition twice the parallax p, and is given in "(seconds of arc). We have the for the distance d = SN: d. p = r, with r the radius E_1S of the orbit of the earth round the sun. When

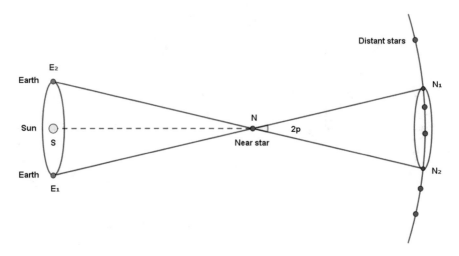

Fig. 1.7 Parallax

p = 1", the distance SN is by definition 1 parsec = 1 pc. Thus: d = 1/p pc. It may be calculated that 1 pc = 3.26 light year (ly). In the figure may be seen that from the earth the near star moves between the distant stars on the celestial globe between the point N_1 and N_2.

Work at the Carte du Ciel was difficult, because the use of a larger telescope was chosen than those of the other observatories. In 1885 in the northern dome on the roof of the Observatory a 10-inch refractor was placed (objective diameter of over 26 cm and focal distance of 4 m). A 7-inch refractor, used till then, was later transported to the Bosscha Observatory in Lembang in the Dutch East Indies.

In 1879 the State Committee for Arc and Level Measurement[7] was set up at the advice of the Academy of Sciences. The aims were the state triangulation and the state height measurements (the precise mapping of The Netherlands); and adaptation of the results to the Middle European measurements. Van de Sande Bakhuyzen was involved from the start and was president of the Committee from 1882 until his death in 1923. He also became secretary of the International Geodetic Association. He was a member of many foreign Academies, for instance associate member of the Royal Astronomical Society in London (De Sitter 1923).

---

[7]Dutch: Rijkscommissie voor Graadmeting en Waterpassing.

Measuring Errors

Accidental Errors

An example is reading errors, which are in equal amounts higher and lower than the correct readings. They may be reduced by doing large amounts of measurements. The errors in the average values are then smaller.

Systematic Errors

One example is the refraction of light rays (for instance from a star) in the successive layers of the atmosphere. As a result of the fact that the temperature and the density of the atmosphere change with height, the speed of light changes as well. That causes refraction and small changes in direction.

A different example is the minute bending of a long telescope by its own weight. The image of a star, in a straight telescope exactly on the cross hairs in front of the eyepiece, is formed slightly besides the cross. If you turn the telescope till the star is on the cross, you read a wrong value.

You can study systematic errors and assess their influence by experiments. The thus found corrections must then be applied to the measurements.

Personal Errors

An example is the determination of the point of time of the passage of a star along a crosswire (estimate between two ticks of the chronometer by punching a hole in a moving paper tape, coupled to the clock). It turns out that one person nearly always punches slightly late and another one slightly early. These errors are hard to estimate. The best observers make the smallest errors.

Ernst Frederik van de Sande Bakhuyzen (1848–1918)

In 1870 Kaiser appointed Ernst van de Sande Bakhuyzen, the ten years younger brother of Hendrik, even before his bachelor exam as second observer in Leiden. Like his older brother he started his studies in Delft and continued them in Leiden. In 1875 he became first observer. His dissertation in 1879 was titled *Determination of the Inclination of the Ecliptic* and was based on declination measurements in Leiden of the sun from the period 1863–1876 (Van de Sande Bakhuyzen 1879).

The Ecliptic
The ecliptic is the plane of the orbit of the earth around the sun. The axis of
rotation of the earth is not perpendicular to the ecliptic. This axis makes an
angle of just over 23° with the direction perpendicular to the ecliptic and is
directed to the Pole Star. This angle causes the succession of the seasons on
the earth.

The observers being active in that period in Leiden, Hermanus Haga (1852–
1936) and Kapteyn, made contributions to these measurements. For the inclination
angle of the ecliptic in 1870 Van de Sande Bakhuyzen found 23°27′22″,18. With
this he had calculated a correction of 0″,13 on the data in the existing tables. Van de
Sande Bakhuyzen was, following his brother, a great analyst of all kinds of errors
which may occur during measurements and reductions. He was according to
observer H. J. Zwiers (1865–1923) in his obituary at the reduction of measurements

the most industrious but also the most scrupulous worker (Zwiers 1918).

With his brother he measured the difference in longitude between Greenwich and
Leiden. He took part in the expedition to Réunion in 1874, led by Oudemans, to
measure the Venus transit. With a large number of other Dutch astronomers Van de
Sande Bakhuyzen was present at the circular solar eclipse in the extreme southern
part of The Netherlands in 1912. After a complex procedure he succeeded his elder
brother as director of the Observatory and stayed in that function till his death. He
became extraordinary professor alongside De Sitter, who became ordinary professor
in theoretical astronomy. De Sitter had a room in the Observatory, with one or two
computers (employees doing prescribed calculations), who worked according to his
orders. But he kept himself far from the policies of Van de Sande Bakhuyzen in the
Observatory. They were friendly with each other. De Sitter let himself be guided by
the idea "if I am not the boss here, I actually don't want to have anything to do with
it". Van de Sande Bakhuyzen's inaugural lecture at the assumption of his
extraordinary professorship was titled *The Meaning the Older Measurements still
have for the Astronomy Nowadays* (Van de Sande Bakhuyzen 1909). De Sitter's
inaugural lecture at the end of 1908 was titled *The New Methods in the Mechanics
of the Celestial Bodies* (De Sitter 1908), in which he broke a lance for the (new
developments in) the old celestial mechanics. There was a remote parallel between
the two inaugural lectures. Van de Sande Bakhuyzen continued the research policies of Kaiser and his brother, although it actually was time to broaden the programme in the direction of the new astronomy. After all great successes had already
been reached with the use of photography, which he did mention in his inaugural
lecture, and spectroscopy. For instance the *Cape Photographic Durchmusterung*
(extensive inspection), the combined project of Gill and Kapteyn. The thousands of
photographic plates of the southern sky, taken at the Cape Observatory at Cape
Town, were measured by Kapteyn in Groningen from 1885 till 1892. It led to a star
catalogue of more than 450,000 stars. Also in the field of spectroscopy results had

been achieved: for instance the composition of the sun and stars. It is true that a photographic telescope (a refractor) had been put up in Leiden in 1898, but till Van de Sande Bakhuyzen's death in 1918 no photos had been made which led to a publication. Internationally the younger Van de Sande Bakhuyzen had less contacts than his brother. He did however take part in the Paris conferences in order to introduce radiographic time signals from the Eiffel Tower.

At his sudden demise the Leiden Observatory had known nearly half a century of flourishing as the largest observatory in The Netherlands in its sixty years of existence: first under Kaiser and later under the elder Van de Sande Bakhuyzen, who brought the Observatory also international fame and influence. The research machine started to falter under the young Van de Sande Bakhuyzen. He chose for the relative safety of the old research programme, he was no innovator. The veneration for Kaiser and his elder brother probably withheld him. That frame of mind was no motivation for colleagues and (doctoral) students to start researches at the Observatory with enthusiasm. Even the presence of the photographic refractor could not induce Van de Sande Bakhuyzen to start a research programme with it. The younger students wanted astrophysics.

Van de Sande Bakhuyzen was an amiable man, with whom it was hard to quarrel. But he was not made of the right stuff to obtain money from the governors of the University and the minister for new investments. Several times he tried to renovate the Observatory, which had become too small. In order to reach that he wrote, when the budget figures had to be turned in, lengthy and always nearly identical reports and petitions year after year, but he never made any headway with his superiors. From the archives one gets sometimes the impression that by a number of colleagues, De Sitter included, and the governors he was not taken completely serious. He must have suffered from it. Perhaps he would have been able as first observer to glory in ever more accurate measurements and reductions. Because he was a sublime observer. He was not known as a gifted orator, but he was loved by many a colleague and (former) pupil. The South African poet Jan F. E. Celliers (1865–1940), who from 1887 studied geodesy in Delft and Leiden, made a poem "In memory of my recently passed away teacher and friend prof. dr. Ernst van de Sande Bakhuyzen, Leiden, Holland" (translation from Afrikaans):

> Modest worker, much have I, sadly,
>
> Forgotten from what your wisdom wished to teach me;
>
> But in my essence your soul is present,
>
> Which is more, and fruitful for still more.

Thus, after Van de Sande Bakhuyzen's sudden demise in 1918, De Sitter, temporarily charged with the directorship, found the Observatory in a situation marked by: few ambitious research programmes, only attention for the old

Fig. 1.8 Ernst F. van de Sande Bakhuyzen. Photograph from the Liber Amicorum at the retirement of Hendrik van de Sande Bakhuyzen in 1908 (in the Leiden Observatory Archive)

positional astronomy, instruments becoming outdated, relatively unmotivated staff, considerable arrears in the reduction and publication of measurements.

The year 1918 was also the year in which the "young" Van de Sande Bakhuyzen turned seventy and would retire according to the law. In the period before his death there certainly must have been discussions in astronomical circles on the succession. Van de Sande Bakhuyzen's death came suddenly, but not his succession.

References

De Jager, C., Van Bueren, H.G., Kuperus, M. (1993). *Bolwerk van de sterren* (Bulwark of the stars), Bekking, Amersfoort, 1993.

De Sitter, W. (1908). *The New Methods in the Mechanics of the Celestial Bodies* (Dutch: *De Nieuwe Methoden in de Mechanica der Hemellichamen*), inaugural lecture, Gebroeders Hoitsema, Groningen, 1908.

De Sitter, W. (1923). 'In Memoriam, Obituary H.G. van de Sande Bakhuyzen'; *RAS, LXXXIV*, 4, February, 1924.

De Sitter, W. (1933). *Short history of the Observatory of the University at Leiden 1633–1933*, Joh. Enschedé and Sons, Haarlem, 1933.

Kaiser, F. (1860). *The Starry Heavens* (Dutch: *De Sterrenhemel*), J.C.A. Sulpke, Amsterdam, J.C. Drabbe, Leiden, 1860.

Kaiser, P.J. (1862). *The application of photography to astronomy* (Dutch: *De toepassing der Photographie op de Sterrekunde*), dissertation, Academic library of P. Engels, Leiden, 1862.

Oort, J. (1928). 'An investigation of the constant of refraction from observations at Leiden and at the Cape'; *BAN*, 4, 143, pp. 137–142, 1928.

Pannekoek, A. (1918). *The Evolution of the Universe* (Dutch: *De Evolutie van het Heelal*), Naaml. Venn. Boekdrukkerij v/h L. van Nifterik Hzn., Leiden, 1918.

Pannekoek, A., Sijes, B.A., Heuvel, E.P.J. van den (1982). *Reminiscences* (Dutch: *Herinneringen*), Van Gennep, Amsterdam, 1982. Note: Pannekoek wrote his *Reminiscences* during World War II in 1944, but did not publish them.

Pyenson, Lewis (1989). *Empire of Reason, Exact Sciences in Indonesia, 1840–1940*; E.J. Brill, Leiden, 1989.

Rooseboom, H. (1994). 'P.J. Kaiser, or: the use of photography in astronomy, 1839–1880' (Dutch: 'P.J. Kaiser of: het gebruik van de fotografie in de sterrenkunde'; *Bulletin of the Rijksmuseum*, 42, 1994, pp. 263–286, 1994.

Sobel, D. (2008). *Longitude*, Harper Perennial, London, 2008.

Spek, H. (1979). *Verificatie van de Rijkszee- en Luchtvaartinstrumenten (1858–1978)(Verification of the Government Sea and Air Force Instruments (1858–1978))*, Marine Electronic and Optical Company (Dutch: Marine Elektronisch en Optisch Bedrijf), 1979.

Van de Sande Bakhuyzen, E.F. (1879). *Determination of the Inclination of the Ecliptic* (Dutch: *Bepaling van de Helling der Eliptica*), dissertation, E.J. Brill, Leiden, 1879.

Van de Sande Bakhuyzen, E.F. (1909). *The Meaning the Older Measurements still have for the Astronomy Nowadays* (Dutch: *De Beteekenis die de Oudere Waarnemingen nog Heden voor de Sterrenkunde hebben*), inaugural lecture, Boekhandel en Drukkerij voorheen E.H. Brill, Leiden, 1909.

Van de Sande Bakhuyzen, H.G. (1863). *On the influence of bending on the heights of celestial lights, measured with the meridian circle.* (Dutch: *Over den Invloed Der Buiging, op de Hoogten van Hemellichten, met den Meridiaan-Cirkel Bepaald*), dissertation, A.H. Bakhuyzen,'s Gravenhage, 1863.

Van Herk, G., Kleibrink, H., Bijleveld, W. (1983). *Leiden Observatory, Four centuries of watching, by day and by night* (Dutch: *De Leidse Sterrewacht, Vier eeuwen wacht bij dag en bij nacht*), Waanders, Zwolle, 1983.

Zuidervaart, H.J. (1999). *Of 'Art Companions' and heavenly phenomena, Dutch Astronomy in the Eighteenth century* (Dutch: *Van 'Konstgenoten' en hemelse fenomenen, Nederlandse Sterrenkunde in de Achttiende eeuw*), pp. 387–389, Erasmus Publishing, 1999.

Zuidervaart, H.J. (2007). *Telescopes from Leiden Observatory and other collections 1656–1859, A descriptive Catalogue*, Rijksmuseum Boerhaave Communication 320, Leiden, 2007.

Zwiers, H.J. (1918). 'E.F. van de Sande Bakhuyzen'; *AN*, 206, p. 161, 1918.

Chapter 2
Descent: Patricians, Lawyers and Politicians

In everything he did a certain casual fearlessness characterized him, perhaps related to the unshaken independence, which revealed in him the descendant of a patrician family.

(Huizinga 1935)

From Flanders, via Leiden and Amsterdam, to Groningen

Willem de Sitter's father, Lamoraal Ulbo de Sitter, descended from a patrician family in Groningen, a university city in the north of The Netherlands. His father, also named Willem de Sitter, published in 1885 a genealogy of the De Sitter family (De Sitter 1885). In this book the family tree was traced back to the beginning of the fourteenth century, to Flanders, now in Belgium. Roughly thirty years later a book from a different member of the family, Mr.[1] J. H. de Sitter, appeared (De Sitter 1916). In that booklet the latter mentioned "besides a lot of good information, also serious inaccuracies" in the genealogy. The official survey of the Dutch patriciate, in The Netherlands the so-called *blue booklet*, follows the data of this J. H. de Sitter. In the blue *Dutch Patriciate*[2], edited since 1910, families are incorporated, when they have uninterruptedly played a prominent role in Dutch society for more than 150 years. He mentions the 16th century Nicolaes (also: Nicolas) de Sitter (also: de Zuttere or de Zittere) as descendant from the West Flemish villages Yperen and Bailleul, close to each other. During skirmishes before the Eighty Years' War (1568–1648) of the Dutch Republic in the making against the Kingdom of Spain, where the Spaniards regularly intervened, this Nicolas de Zittere was summoned before the bailiff of Bailleul in 1567. Thereupon he fled to Antwerp with his son, also named Nicolaes, where he became a police master. After the overthrow of Antwerp in 1585 he left for Leiden. He was a fortunate man, married three times and died there in 1601. From the register of deaths in Leiden it shows that his father was named Willem.

---

[1]In The Netherlands the title *Mr.* is awarded to anybody passing the master's exam of the study of law.
[2]Dutch: Nederland's Patriciaat.

© Springer Nature Switzerland AG 2018
J. Guichelaar, *Willem de Sitter*, Springer Biographies,
https://doi.org/10.1007/978-3-319-98337-0_2

His son Nicolaes stayed in Leiden only for a short time and went to Amsterdam. In the Old Church[3] there he married Barbara Willekens. There a son, a third Nicolaes, was baptized in 1596. The second Nicolaes chose an adventurous career and signed up with the Dutch East India Company as chief merchant on the sailing vessel Leyden, which left the roadstead of the island of Texel for Dutch East India in 1600, together with five other vessels under the command of admiral Van Neck. There it seems Nicolaes died in an incident near Sumatra. These De Sitters were adventurous, ambitious and not afraid.

The third Nicolaes was a gilder in the Warmoes Street (street with vegetable sellers) in one of the oldest parts of Amsterdam. He married Sara Cornelis and they had two sons, Frans (wine merchant, born in 1631) and Pieter (silk retailer, born in 1636). The two brothers married in 1663 the sisters Janneke and Aeghje, daughters of Willem Cracht, shopkeeper in silk and cloth. Pieter was sworn in as a burgher and with his wife Aeghje he had six children. The third son was Wilhelmus, baptized on 13 September 1671 in the Northern Church[4] in Amsterdam. Later he went to Groningen and there he married Lubbina Johanna Hendrica Princen. Not a lot is known about Wilhelmus/Willem. Certain is that he was part of the bourgeoisie.

Wolter Reinolt, Primogenitor of Lawyers and Civil Servants

Wilhelmus de Sitter and his wife Lubbina Princen had a son Wolter Reinolt in 1709. With him the De Sitter family of lawyers, politicians and civil servants started. Wolter Reinolt studied law from 1727 at the University of Groningen (the name of this university changed several times in the course of its history since 1614; we only use the word university) and wrote his dissertation in Latin about *Emphyteusis* or ground lease. He became a judicial civil servant in 1737. The Republic of the United Netherlands was governed by a so-called stadtholder. During two stadtholderless periods (1650–1672 and 1702–1747) no stadtholder was appointed. After the second period Willem IV became stadtholder of all provinces and Wolter Reinolt, supporter of the Orange party, was appointed Secretary of the city of Groningen in 1749. That was a raw deal for his predecessor, a certain Herman van Gesseler, who in blind rage and unashamedly tried to blacken De Sitter's good name:

> Very naughty bloke, one of the most heinous mutineers and most godless rioters, a big debauchee and waster, negligent and careless, adulterer and whore hunter, assaulter and skinner of all he can put his claws into.

---

[3]Dutch: Oude Kerk.
[4]Dutch: Noorderkerk.

It was not easy to enter the higher circles of civil servants and local government. This forefather of the lawyers family De Sitter must have developed a hide of a rhinoceros by all these insults. But his career was rising. From 1764 he was a judge, deputy to the States General[5] of the Republic and member of the Provincial Executive[6] of the province of Groningen. He set the tone for the lawyers after him.

Willem de Sitter (1750–1827)

His son Willem, the great-great-grandfather of our astronomer, studied in Groningen and received his doctor's degree there with his dissertation in Latin *De fictionibus iuris patrii aeque ac romani* (About the fictions of Dutch and Roman law) in 1772. Then a beautiful legal career followed. A number of his posts were: barrister, member of the High Court of Justice of Groningen, bailiff of Westerwolde and president of the court of law in Winschoten (a region and a town in the province of Groningen). One of the children from his marriage with Maria Albertina Johanna de Drews (1756–1828) was Johan.

Johan de Sitter (1793–1855), Soldier

Johan de Sitter, great-grandfather of the astronomer, studied law in Groningen and received his doctor's degree on the basis of a number of propositions in 1815. He served from 1812 till 1823 as a lieutenant in the National Guard and in the infantry, in which function he took part in the siege of Delfzijl (1813–1814), a town in the northern part of the province of Groningen, in order to drive the French out of the town. He continued his career as a controller and an inspector of the tax authorities. In 1830–1831 he was in active service as a captain and a major in the Frisian Militia. In August 1831 he took part in the Ten-Day Campaign under King William I in the secessionist Southern Netherlands, which had proclaimed the monarchy under King Leopold after the Belgian uprising. Johan received the Military Order of William I for his brave service. Then he became a notary in Groningen and from 1850 till his death in 1855 he was as a liberal member of the Dutch Senate, specialized in water management, internal affairs and finance. Johan had married Anna Antoinetta countess Van Limburg Stirum (1792–1863) in 1817. The De Sitter family had already taken their place in the Dutch patriciate, and now were also married into the Dutch nobility. Johan and Anna had two children: Maria Albertina Johanna and Willem.

---

[5]Dutch: Staten-Generaal.
[6]Dutch: Gedeputeerde Staten.

Fig. 2.1 Family coat of arms.Coat of arms of the De Sitters

Fig. 2.2 Willem de Sitter's great-grandparents Johan de Sitter and Anna Antoinetta countess Van Limburg Stirum. Photographs from the family photo album of the astronomer's brother Ernst Karel Johan, now in the possession of his grandson and namesake Ernst K. J. de Sitter

Willem de Sitter (1820–1889), Mayor of Groningen

This Willem, grandfather of our astronomer, attended grammar school in Groningen and studied law there. He took his doctor's degree in 1843. Besides his practice as a barrister he held as a liberal a large number of political and executive functions. As a political representative he was a member of the City Council[7] of Groningen, a member of the Provincial States[8] and from 1877 till 1888 nearly uninterruptedly a member of the Dutch Senate.[9] As a politician he was an alderman and mayor of the city of Groningen and member of the Provincial Executive. His functions also included substitute judge and chief governor of the University. For his great merits he became Knight of the Order of the Dutch Lion, Knight of the Order of the Pole Star (a Swedish decoration) and Commander in the Order of the Oak Crown (an original Luxembourg decoration, which was often awarded by King William III). Willem married the noble Wilhelmina Petronella baroness Rengers (1825–1895), daughter of Lamoraal Ulbo baron Rengers and Catharina Louise van Naerssen.

---

[7]Dutch: Gemeenteraad.

[8]Dutch: Provinciale Staten.

[9]Dutch: Eerste Kamer.

Fig. 2.3 The paternal grandparents of Willem de Sitter: Willem de Sitter and Wilhelmina Petronella, baroness Rengers. Photographs taken from the family photo album of the astronomer's brother Ernst Karel Johan

Since then the first names Lamoraal Ulbo frequently appear in the De Sitter family. They had four children: Johan, Anna Antoinetta, Wilhelmina Petronella and Lamoraal Ulbo. Grandfather Willem had not only a certain feeling for standing, but also for economy. He took part in a tour through the new building of the Department of Justice in 1880. The building costs were estimated at 200,000 Dutch guilders, while in the end they were nearly 900,000 guilders. Also in those days estimates of large projects were made poorly. In a kitchen De Sitter noticed a number of tiles painted by the famous Dutch painter Van de Velde. In his opinion they did not belong in the kitchen of a janitor, but in a museum (Verburg 1994).

Compact Family Tree of Willem de Sitter, the Astronomer

- I—Willem, 16th century. Flanders?
- II—Nicolaes de Sitter (I), de Zittere, de Zuttere. Died in Leiden in 1601.
- III—Nicolaes de Sitter (II). Leiden, Amsterdam. Married to Barbara Willekens.
- IV—Nicolaess de Sitter (III). Born in Amsterdam in 1596. Married to Sara Cornelis. Sons Frans and Pieter.

- V—Pieter de Sitter. Born in Amsterdam in 1636. Married to Aeghje Cracht in 1663. Six children.
- VI—Wilhelmus de Sitter. Born in Amsterdam in 1671. Groningen. Married to Lubbina Johanna Hendrica Princen. Two of their children were daughter Agatha and son Wolter Reinolt.
- VII—Wolter Reinolt de Sitter (1709–1780). Married to Johanna Schultens in 1742. Two of their children were the sons Albert Johan and Willem.
- VIII—Willem de Sitter (1750–1827). Married to Maria Albertina Johanna de Drews (1756–1828). Son Johan.
- IX—Johan de Sitter (1793–1855). Married to Anna Antoinetta countess Van Limburg Stirum (1792–1863) in 1817. Two children: Maria Albertina and Willem.
- X—Willem de Sitter (1820–1889). Married to Wilhelmina Petronella baroness Rengers (1823–1895) in 1843. Four children: Johan, Anna Antoinetta, Wilhelmina Petronella and Lamoraal Ulbo.
- XI—Lamoraal Ulbo de Sitter (1846–1908). Married to Catharina Theodora Wilhelmina Bertling (1846–1909) in 1871. Three children: Willem, Wobbine Catharine (1876) and Ernst Karel Johan (1879).
- XII—Willem de Sitter (1872–1934). Married to Eleonora Suermondt (1870–1952). Five children: Lamoraal Ulbo (1899–1901), Theodora (1900), Lamoraal Ulbo (1902), Aernout (1905) and Agnes (1908).

References

De Sitter, W. (1885). *Genealogy of the De Sitter Family* (Dutch: *Genealogie van het Geslacht De Sitter*), J.B. Wolters, Groningen, 1885.

De Sitter, Mr. J.H. (1916). *Genealogical notes on the De Sitter Family* (Dutch: *Genealogische aantekeningen over het geslacht De Sitter*), H. ten Brink, Arnhem, 1916.

Huizinga, J. (1935). 'In memoriam W. de Sitter'; *Almanac of the Leiden Students' Union* (Dutch: *Almanak van het Leidsche Studentencorps*), 1935.

Verburg, Dr. Mr. M.E. (1994). *History of the Ministry of Justice, part I, 1798–1898* (Dutch: *Geschiedenis van het Ministerie van Justitie*), Sdu Publishers, The Hague, 1994.

Chapter 3
Youth and Education

Willem is untidy at mathematics.
Staff meeting of the City Grammar School of Arnhem in 1885–
1886. (In the Archives of the province of Gelderland)

The De Sitter Family

Willem de Sitter's father, Lamoraal Ulbo, was born on 5 mei 1846 in Groningen and was named after his grandfather on the maternal side. After his studies he received his doctor's title with a dissertation titled *The printing press as a means for crime* (De Sitter 1869). It was a major work of more than 200 pages with no less than 42 extra propositions. On his dissertation a positive review appeared:

> Because improvements in the legislation cannot be prepared by sitting around. And that is why the treatise of Mr. De Sitter may hopefully obtain many readers in our legal public! (Mom Visch 1869)

Then he became a barrister at the provincial court in Groningen for some time. He also worked as a civil servant of the prosecution in Assen (in the province of Drenthe). In March 1871 De Sitter was appointed as a deputy clerk of the district court in Sneek (in the province of Friesland) and he moved to Sneek in April 1871.[1] He was then nearly 25 years of age and in that same year he married Catharine Theodore Wilhelmine Bertling (1846–1909) from Assen, a daughter of court clerk Ernst Karel Johannes Bertling and Wobbina Catharina van der Scheer.[2] Already in 1872 Lamoraal Ulbo was appointed judge. The married couple moved into the northern half of a double mansion, 47 Water Gate Canal,[3] with a nice view of the Water Gate (Ten Hoeve 1999). Lamoraal Ulbo de Sitter celebrated his 26th birthday on 5 May 1872, but the

---

[1]*Municipal register of Sneek* (Dutch: Bevolkingsregister); Town hall in Sneek.
[2]*Municipal register of Assen*; Archives of the province of Drenthe in Assen (Dutch: Drents Archief).
[3]Dutch: Waterpoortsgracht.

© Springer Nature Switzerland AG 2018
J. Guichelaar, *Willem de Sitter*, Springer Biographies,
https://doi.org/10.1007/978-3-319-98337-0_3

Oud 30 jaar

Fig. 3.1 Willem de Sitter's parents Lamoraal Ulbo de Sitter and Catharine Theodore Wilhelmine Bertling. Photographs from the family photo album of the astronomer's brother

celebration was not exuberant, because his wife was probably already in labour and during the night, just after two o'clock a son was born. The next morning the judge made the registration. The son was given a single first name: Willem. As was the custom he was named after his paternal grandfather. The judge had taken two respectable witnesses to the registration's office: Hendrik Albarda and Jan Telting, a clerk and a deputy clerk from Sneek.[4]

The De Sitters left for Rotterdam in 1875, where Lamoraal Ulbo had been appointed judge at the district court. Those years were turbulent in the whole of the Dutch judiciary due to an extensive reorganisation. Six provincial courts were abolished. Only Leeuwarden (in Friesland), Arnhem (Gelderland), Amsterdam (North-Holland), The Hague (South-Holland) and Den Bosch (North-Brabant) remained. In 1876 all justices at these courts lost their jobs. They had to apply again, creating the possibility not to reappoint *difficult boys*. The same fate struck all the judges at the lower courts in 1877.[5] As a result of this *migration* Lamoraal Ulbo was appointed in Arnhem. After moving to Arnhem the De Sitter family moved another two times, the last time to 106 Boulevard Heuvelink in 1897–1898.[6]

---

[4]*Municipal register of Sneek.*

[5]Information about the reshuffling of the courts was given to me by Mr. Van Loo, judge at the court in Arnhem, in a private communication in 2007.

[6]*Municipal register of Arnhem*; Archives of the province of Gelderland in Arnhem.

Willem's Sister Wobbine and Brother Ernst Karel Johan

A second child was born on 8 August 1876, a girl: Wobbine Catharine. The family were staying at the grandparents in Assen. Lamoraal Ulbo was still a judge in Rotterdam. Wobbine attended the Secondary School for Girls, together with her girlfriend the Dutch painter Mies Elout. They regularly played tennis with Willem. The Elout family moved to the seaside resort Domburg in the province of Zeeland, where Wobbine was a regular guest. After the death of her mother in 1909 (her father had died in 1908) Wobbine went to live in Domburg. Later she went to Lausanne in Switzerland for a training to become a nurse. The famous Dutch painter Piet Mondriaan lived temporarily in her house in that period. Wobbine was part of the artists' circle round the painters Jan Toorop and Piet Mondriaan, and the author Arthur van Schendel, who met frequently in Domburg in the years 1910–1920. Wobbine received a few paintings from Mondriaan as presents. She also knew Johan Huizinga, who had become a good friend of her brother Willem during their studies in Groningen. He became one of the most prominent historians of The Netherlands. Huizinga had become a widower in 1914 and was appointed professor in Leiden shortly thereafter. For a short period in 1915 the Huizingas stayed at the large house of the De Sitters in the Observatory, before moving to a new house. To thank the De Sitter family for their hospitality they invited the De Sitter family to a joyful day of sailing

> to celebrate the liberation of their territory from the invasion of the Huizingas.

> (Van der Lem 1993)

In this difficult period for Huizinga Wobbine supported him. She worked till 1917 as a housekeeper for Huizinga. The author of a lot of picaresque novels Leonhard Huizinga, son of Johan, wrote in his *Recollections to my father* about her:

> The woman to whom we children owe immortal gratitude. Aunt Bine, Bine de Sitter, the sister of his old friend De Sitter, who in many respects has been a mother to us in the fullness of its meaning for years. (Huizinga 1963)

But she had to leave, because Johan Huizinga was afraid that due to her his children would forget their own mother. At the end of World War II her house in Domburg was destroyed by a direct hit in a bombardment during the battle of the Schelde. At that moment Wobbine lived in Oegstgeest temporarily, near Leiden. Later she had a new house built in Domburg. She died in 1958. Her brother Willem spent many a holiday in Domburg, perhaps also because of the supposedly healthy sea air, which would be wholesome for his weak lungs. Many members of the family regularly stayed with aunt Bine, Bienk or Bienkie (depending on age and part of the family different names were in use). All of them had the best recollections of their visits.

On 12 November 1879 Willem and Wobbine got a brother in Arnhem, Ernst Karel Johan.[7] Later Ernst would reckon his visits with the family to his grandparents De Sitter, who had a countryhouse in Dieren, near Arnhem, to his most pleasant recollections. From Arnhem they first took the horsetram to Velp and from there the steam tram to Dieren.[8] Ernst made his career in a financial institution and was very interested in new technological developments. After a visit to his younger brother, who then lived outside Groningen, Willem de Sitter spoke admiringly about the new gadget of his brother: an automatic central heating system, fired with oil from a tank in the garden and with a thermostat in the living room, which you could set to a certain temperature. Ernst died in 1949.

Willem, his parents, brother and sister had close family ties, also after his marriage to Eleonora Suermondt. In the book she wrote about her husband after his death we read:

> When they [the children of Eleonora and Willem] were still small, we spent our holidays in the delightful parental house in Arnhem. (De Sitter-Suermondt 1948)

Willem loved making long walks on the Veluwe (a large nature reserve in the province of Gelderland). From his youth he knew the area well. His father wrote him in the beginning of 1907 after reception of an article about the satellites of Jupiter from *The Nature*[9] (a popular illustrated monthly, dedicated to the sciences), which he called a thorough piece of work. He also wrote that in a so positive science as that of Willem a lot of work was done on the basis of probabilities. Moreover, he advised him not to take too much on his plate. He did not write about family details, those Willem would probably hear from his mother and Bine:

> Many warm regards from the three of us, give the children a kiss from your loving father, L. U. de Sitter.[10]

And when Willem de Sitter received the message of his appointment as professor in Leiden in 1908, when he stayed with wife and children in Bergen (in the province of North-Holland), his parents, sister and brother came there to share his happiness (De Sitter-Suermondt 1948).

---

[7]*Municipal register of Arnhem.*

[8]The data come from an annotated photo album of Willem de Sitter's brother Ernst, which is now in the possession of his grandson Ernst Karel Johan de Sitter.

[9]Dutch: De Natuur.

[10]De Sitter, L.U., letter of 24 February 1907, in possession of De Sitter's granddaughter Tjada van den Eelaart-de Sitter.

Fig. 3.2 Willem and his sister Wobbine (or Bine) and brother Ernst in 1891. Photograph from the family photo album of the astronomer's brother

Career of Father Lamoraal Ulbo de Sitter

Father De Sitter declined a professorship in 1884. For the rest of his career he stayed at the judiciary in Arnhem. He switched to the higher court in Arnhem, where he became judge and president in 1897.[11] In that year the De Sitter family made their first trip abroad, to Remouchamps and Spa in Belgium. His career as a judge was excellent, be it not so spectacular as those of a few of his forefathers, who alternated or combined their work in the judiciary with high posts in politics at local, regional and national level. A few times he tried to make a further move in his career. In the years 1885–1895 his name was on the list of recommendations for the Supreme Court[12] a number of times, but it never came to an appointment (Van Koppen and Ten Kate 2003). But he did collect board posts: Member of the Civil Poor Relief, Member of the Court of Appeal, Member of the Civil Hospital, Governor of the Dullert Foundation (named after W. H. Dullert; the foundation

---

[11]Information from Mr. Van Loo, judge at the court in Arnhem, in a private communication in 2007.

[12]Dutch: Hoge Raad.

helped poorer inhabitants of Arnhem), Governor of the City Grammar School of
Arnhem. For all his work he was honoured with a royal decoration: Knight in the
Order of the Dutch Lion. It was said about him that his word had authority in all
circles and everywhere in the country.[13] He died on 6 September 1908, at the age of
62, bequeathing a generous inheritance to his wife and children.[14] At his funeral
Esquire Rethaan Macaré spoke:

> A paragon of a noble, truthful and highly educated man, who showed a happy union of gifts
> as are only given to few. (Photo album of E.K.J. de Sitter.)

His wife Catharine moved to Groningen, where she died in 1909.

Willem's Time at the City Grammar School

There is not much documentation on Willem's early youth. In 1885 he went to the
grammar school, and not to the Higher Civilian School[15] (hbs), which had been
founded in Arnhem in 1866. The economic growth in The Netherlands after 1850
made a good education for civilian jobs ever more necessary. Secondary education
aimed   at   general   civilian   professions   was   needed.   The   Dutch   Prime
Minister J. R. Thorbecke (1798–1872) submitted an education law in 1863, which
passed parliament. The law made it possible to found new hbs's, some with a three
year course and many with the very successful five year course. Grammar schools
stayed part of the higher education, preparing for university. The middle classes
could attend the hbs without right to enter university, the higher classes to the
grammar school and the university. In spite of, or perhaps due to this, in our
present-day view, remarkable distinction the hbs instantly became a resounding
success. Many hbs pupils could enter university after an additional exam in the
classical languages. In particular the hbs pupils of the B-side (with emphasis on
mathematics and the sciences, whereas the A-side had emphasis on the modern
languages and accounting) did extremely well. It yielded The Netherlands a number
of Nobel prize winners: J. H. van 't Hoff, H. A. Lorentz, P. Zeeman, H. Kamerlingh
Onnes and W. Einthoven (Willink 1998). Long before the introduction of the hbs
the Latin School and the City Boarding School in Arnhem merged into The City
Grammar School in 1842.[16] One department of the school was a grammar school,

---

[13]The data come from an annotated photo album of Willem de Sitter's brother Ernst, which is now
in the possession of his grandson Ernst Karel Johan de Sitter.

[14]*Memorandum of inheritance of Lamoraal Ulbo de Sitter* (Dutch: *Memorie van aangifte der
nalatenschap*); Archives of the province of Gelderland in Arnhem.

[15]Dutch: Hogere Burger School (hbs).

[16]Inventory and documents of the archive of the Board of Governors of the City Grammar School
in Arnhem, 1655–1970, and of the archive of the City Grammar School itself, 1657–1976, all in
the Archives of the province of Gelderland. A lot of data about Willem's time at grammar school
can be found in these archives.

preparing the pupils for an academic career, the teaching in the other department (with mathematics, sciences and modern languages) was aimed at civilian jobs immediately after the school exam. This part was more or less a school like the hbs of twenty years later. However, the merger was a failure and after twelve years the schools went their own ways again: The City Grammar School and The City Institute.[17] At the Grammar School mathematics, sciences, modern languages, history and geography were taught from then on. However, it went through a difficult period. In the school year 1878–1879 there were only 53 pupils.

When Willem had to go to a secondary school, the new hbs in Arnhem had already been in existence since 1866, but in the judicial administrative De Sitter family grammar school was the appropriate school for Willem. In the school guide of the year 1882–1883 father De Sitter is mentioned as a governor of the City Grammar School, in 1885–1886 he was secretary and later he became president of the Board of Governors. In 1885 Willem was enrolled at the Grammar School of dr. F. Goslings, rector from 1879 till 1913. Willem's name appears in the *Matricula*, a book in which all pupils since 1657 had written their names, places and dates of birth in Latin: *Willem de Sitter, Snecanus, natus die 6 m Maii a 1872*. In that year the school counted 78 pupils, the rector, eleven teachers, five of them with a doctor's title, and the *claviger* (Latin for "carrier of the keys"; caretaker). In the first year he had "only" 28 lessons a week in 7 subjects, 8 in Latin and 4 in mathematics. The complete scheme for all years in the school is given below:

Subject/Year	1	2	3	4	5α	5β	6α	6β
Greek	0	6	6	7	7	5	8	5
Latin	8	6	6	6	7	5	7	4
Dutch	3	2	2	2	2	2	1	1
French	4	2	2	2	1	1	1	1
German	0	3	2	2	2	2	1	1
English	0	0	3	3	2	2	1	1
History	4	3	3	2	2	2	3	3
Geography	3	2	2	1	0	0	1	1
Mathematics	4	3	3	3	3	5	2	5
Sciences	0	0	0	0	2	2	2	2
Biology	2	2	0	0	0	2	0	2
Total	28	29	29	28	28	28	27	26

A few things are conspicuous: the great emphasis on the classical languages, the few lessons in the mother tongue, the relevant number of lessons in history, the English language started only in year 3, and the small number of lessons in the sciences (physics and chemistry) and biology in the β department (years 5 and 6). Latin, Greek and mathematics were the prominent subjects in this department. The

---

[17]Dutch: Het Stedelijk Instituut.

total number of weekly lessons was about 10% less than in the present secondary schools.

The school guide stressed two articles from the *Statute for the regulation of the Grammar School in Arnhem.*[18] Article 23 about the admission conditions to the first year: at least 12 years (Willem is 13) and passing the entrance exam on the subjects: reading, writing, arithmetic, the basics of the Dutch and French language, and of history and geography. Further Article 28 about neglect of duty or misconduct. The possible punishments were: writing lines, dismissal from the classroom, a maximum of eight days expulsion from school (the rector had to inform the trustees). Only the trustees could expel a pupil indefinitely (except for appeal to mayor and aldermen). Probably it was necessary to make clear that misconduct was not tolerated.

In the reports of the teachers meetings Willem is hardly mentioned. The only remark found dates from his first year: "Willem is untidy at mathematics". On his first school report he had nearly only marks between 4 and 5 (on the then used scale from 1 till 5). Only his mark for "behaviour" at history and geography was 2. Still he seems to have been a quiet pupil. On 2, 3 and 4 July 1891 Willem did his final exam, in both departments: α and β. His two certificates gave him access to all university faculties. With a certificate α you could only enter the faculties of theology, law or literature and philosophy. The certificate β gave only access to the faculties of medicine or mathematics and physics. Three of the eleven pupils in his class got the certificates α and β, six only α and two only β. The average mark of Willem was 4. He was examined in Greek ($3\frac{1}{2}$), Latin ($4\frac{1}{2}$), Dutch and history ($3^2/_3$), modern languages ($4^1/_3$) and mathematics (4). Indeed, an excellent pupil. He had the maximum 5 for English and one of the partial subjects of mathematics.[19]

Study in Groningen

In the official *Album Studiosorum Academiae Groninganae*, the list of names of the University of Groningen, Willem's name is enlisted on 21 September 1891 for the department of mathematics and physics. On 17 October Johan Huizinga is enlisted for the department of literature. It was the start of a lifelong friendship. Willem became a member of the Students' Union of Groningen, *Vindicat Atque Polit* (Maintain and Refine). It was founded in 1815 and the oldest students' union in The Netherlands. The Students' Almanac of Groningen, which appeared yearly, gives some information about Willem's student days. The Union had 271 members in 1891, while there were 230 non-member university students. The faculty of

---

[18]Dutch: Verordening tot regeling van het Gymnasium te Arnhem.

[19]The list with Willem's exam marks is present in the school archive of the City Grammar School in Arnhem.

Fig. 3.3 Willem de Sitter at 16 years of age. Photograph from the archive of De Sitter's grandson Willem Jan de Sitter

mathematics and physics counted 67 students. The whole university had less students than an average primary or secondary school a century later. The total of university students and that of the faculty of mathematics and physics would remain more or less constant during Willems period at university. In the Almanac of 1893 the lectures on astronomy of the probably popular professor Kapteyn were mentioned twice:

> The lectures on astronomy of professor Kapteyn were also very interesting this year.

> We rejoice in advance at the professor's lectures on astro-physics.

From the Almanacs it is not clear if Willem was a regular visitor of the Union's building *Mutua Fides* (Mutual Trust), although, according to Huizinga, his studies did not prevent Willem from devoting himself with appetite to the joys of the student life (Huizinga 1935). Certainly he was a very serious student. In the Almanac of 1894 Willem was mentioned as secretary of one of the subsocieties of the Union: *Collegium Nationale Gelriae ac Transisalania*, founded in 1879 for students from the provinces of Gelderland and Overijssel. The members assembled at the digs of one of the members on Sundays at 15:00 h. The Almanac of 1894 was again positive on the lectures of Kapteyn, and it mentioned the bachelor's exams of Willem de Sitter and Johan Huizinga, on 9 July and 29 October 1894. He passed his bachelor's exam within three years. Willem was certainly not on his way to become an "eternal" student. He was a keen billiards player and was chosen in the Billiards Committee after a hot ballot battle, as mentioned in the Union's bulletin *Fata of the Union*. After the ballot they played a 25 hits billiards game.

The Maskerade was a pageant with a historical subject, organized by the students every fifth anniversary of the University since the founding of the Union. In 1892, for the Maskerade in 1894, the theme chosen was the entry of Prince Maurits, the son of William of Orange, in Groningen in 1594. In the Eighty Years' War against the Spanish Groningen had chosen for the Spanish king Philips II. After a siege of the city the princes Maurits and Willem Lodewijk could enter the city and bring it back into the Republic. Johan Huizinga was a member of the Maskerade Committee and Willem rode on horseback in the parade.[20]

Much later, Huizinga would describe Willem in his student years in his obituary (Huizinga 1935):

> A long, slender figure with a fine face, already somewhat pale, under fair hair, which had not yet completely lost the curls from his boyhood. His somewhat clumsy movements did slightly harm his well-formed shape. He was a good horseman and an excellent skater.
>
> The dominant factor in his appearance were his eyes: big dark blue eyes, a steady and calm look, without secrecy or agitation. He spoke little, but took part in a lively way in what others said, smoking his pipe, till he suddenly took the floor with enthusiasm on a subject that interested him.

Willem not only occupied himself with mathematics, but also with philosophical thoughts. During a stay in Baden-Baden in Germany in 1896 he tried to put his thoughts on paper: about the way in which we perceive reality. In his original hand-written text there are some words that are difficult to read. These have been supplied with a question mark between brackets [ASL, Note from 1896].

> A physical theory (?) for example electricity is not an "explanation" of the as in reality existing facts to be accepted, but a conformal representation. All our perceptions are conformal representations. Thus, thinking is the formulation and elaboration (or in fact only observation) of conformal representations; only by making a conformal representation of something we are able to understand that something we name. If we wish to understand our own thinking, we must have a conformal representation of it, but in what? If it is possible in something, then in the abstractly mathematical. That is not the abstract conception of space or magnitude or something, those are only things that can also later (?) be represented in the primal-abstract-real-mathematical (and then become mathematics), but that deepest, single, real existence is something that we do not know, that we cannot know, of course, because in what would we represent "that"? It is not only the quintessence of mathematics but of logic, of the symbolism, of the truth in art, of the faith of the feeling, and it must be possible by subtle

---

[20]Much information about Willem de Sitter's student years is to be found in the archives of the University Museum in Groningen.

Fig. 3.4 Drawing of Willem de Sitter as a student by his friend Huizinga. Translation of the text on the drawing: Say, what would life be without mathematics? a world without a sun/yes, a shadow which nearly vanishes. drawn by H. Huizinga. this is Willem de Sitter/24 years of age/ mathematical artist. Drawing from the family archive

reasoning, demanding strong intellectual powers, to represent all these things in a certain minimum of abstract symbols, which then represent that real thing, (but?) are not the real self.

Baden-Baden, 10 August 1896.

He tried to represent the thinking and the reality on mathematical notions. He may have read Hertz, who said that scientific theories are images of reality. Or we could make a parallel with Plato's cave, in which the prisoners can only see before them the shadows of the reality behind them. But perhaps these were Willem's original thoughts.

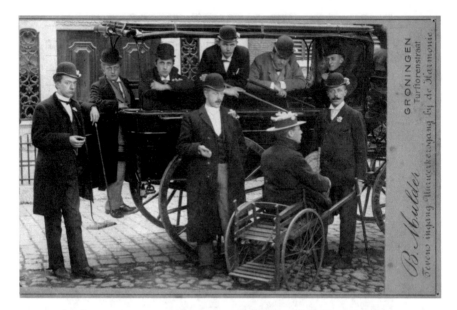

Fig. 3.5 An exam party in 1893 in Groningen. Willem in the carriage (top middle, with two fingers on his walking stick). On his right hand his lifelong friend and later famous historian Johan Huizinga. Photograph from the University Museum in Groningen

How Astronomy Came into De Sitter's Life

As a child Willem was already fascinated by the stars. In the garden in Arnhem he put together a small telescope. He mused on the structure of the universe and made phantasies about a complete different earth (De Sitter-Suermondt 1948). During his time at grammar school his interest in the sciences must have developed.

An obvious question is: why did Willem choose for a career as a professional astronomer in the end? His talent for mathematics and the sciences, but also the broad interests of his father are elements in the explanation. A selling catalogue from 1909 of books from the inheritance of his father and two other men shows: *The Starry Heavens*[21] of Frederik Kaiser (predecessor of Hendrik van de Sande Bakhuyzen and builder of the new Leiden Observatory in 1860), and *The wonders of the heavens*[22] by Camille Flammarion. For a young man with a growing interest in astronomy Kaiser's book must have been a revelation.

---

[21]Dutch: *De Sterrenhemel*.

[22]Dutch: *De Wonderen des Hemels*. The catalogue was published by S. Gouda Quint: *Catalogue of Books, Pictures* etc., partly left by the gentlemen Mr. L. U. de Sitter, President of the Court, Mr. Th. Van Tricht, Dr. A. J. van Rossum; Arnhem, 1909. So it is not certain that these books came from the library of De Sitter's father.

In the De Sitter family it was absolutely not a matter of course that the young Willem took up his studies in mathematics and physics. The most obvious choice when entering Groningen University in 1891 was the study of law, as a preparation for a career in the judicial field or a higher function in the administrative field. So it had been for about two centuries from father to son.

Willem's talent and love for mathematics and the sciences had shown clearly during his education at the City Grammar School in Arnhem. In Arnhem there was also an hbs, but this gave no direct access to a university education. A university study in mathematics and physics was very probably Willem's own choice. Mathematics was his first option. If his parents immediately consented, is not known. A fact is that his parents were very proud of their successful scientific son and supported him at the start of his astronomical career. First with the two year period in Cape Town under the guidance of Sir David Gill and later when he was an assistant of Kapteyn between his doctorate in 1901 and his appointment in Leiden in 1908. Later De Sitter would write that his father was more proud of his appointment as a professor than he himself and that it was a pity that he had not been able to be present at his inaugural lecture. He died shortly after De Sitter's appointment.

After his bachelor's degree in 1894 he obtained a small job as calculator at the Astronomical Laboratory of Kapteyn. Perhaps the very lectures of Kapteyn, very popular with all the students, had done their magical work already, so that De Sitter deliberately chose for the job with Kapteyn. In the De Sitter Archive two lecture notebooks are present, made during the lectures by Kapteyn: Mechanics II and III. Part II dates from the year 1894–95, immediately after his bachelor's exam. The lectures dealt, among other subjects, with Kepler's laws, for De Sitter his first serious acquaintance with celestial mechanics.

At the end of 1896 David Gill, director of the Observatory in Cape Town, visited his good friend Jacobus Cornelius Kapteyn for a few days. Between 1885 and 1892 Kapteyn had reduced[23] for Gill his photographic observations of the Southern Hemisphere: the measurement and calculation of the data of all the stars visible on thousands of photographic plates. At the invitation of Gill (probably a joint plan with Kapteyn) De Sitter went to Cape Town to work for a period of two years (1897–1899) at his Observatory. There Gill implanted in De Sitter the real love for astronomy. His stay in South Africa was decisive for his further career. There he also met a second love: Eleonora Suermondt, whom he married in 1898. De Sitter would always love South Africa and went back a few times.

So Gill's visit to Kapteyn in 1896 made a complete mess of Willem's well-organized study plans in mathematics.

---

[23]'Reducing' means the process of calculating relevant astronomical data from a mass of measurements.

References

De Sitter-Suermondt, E. (1948). *Willem de Sitter, Life of a man* (Dutch: *Een menschenleven*), H.D. Tjeenk Willink & Zoon N.V., Haarlem, 1948. A first edition of the book appeared in 1940. In 1948 a chapter about their trip to America was added.

De Sitter, L.U. (1869). *The printing press as a means to crime* (Dutch: *De drukpers als middel tot misdrijf*), Wolters, Groningen, 1869.

Huizinga, J. (1935). 'In memoriam W. de Sitter'; *Almanac of the Leiden Students' Union* (Dutch: *Almanak van het Leidsche Studentencorps*), 1935.

Huizinga, L. (1963). *Recollections to my father* (Dutch: *Herinneringen aan mijn vader*), H. P. Leopold, Den Haag, 1963.

Mom Visch, D.J. (1869). *Juridical Magazine* (Dutch: *Regtskundig Tijdschrift*), 4th Piece, 1869.

Ten Hoeve, Sytse (1999). *Sneek in photographs 1863–1900* (Dutch: *Sneek gephotographeerd*), Friese Pers Boekerij, Leeuwarden/Ljouwert, 1999.

Van der Lem, A. (1993). *Johan Huizinga, Life and work in pictures and documents* (Dutch: *Leven en werk in beelden & documenten*), Wereldbibliotheek, Amsterdam, 1993.

Van Koppen, P.J. and Ten Kate, J. (2003). *Appointments to the Supreme Court of The Netherlands 1838–2002* (Dutch: *Benoemingen in de Hoge Raad der Nederlanden*), Kluwer, Deventer, 2003.

Willink, B. (1998). *The Second Golden Age, The Netherlands and the Nobel prizes for the natural sciences 1870–1940* (Dutch: *De Tweede Gouden Eeuw, Nederland en de Nobelprijzen voor natuurwetenschappen 1870–1940*), Bert Bakker, Amsterdam, 1998.

Chapter 4
The Making of an Astronomer

The man who never makes a mistake is he who does nothing.
David Gill (Kapteyn 1914)

Short History of the Royal Observatory of Cape Town

In the beginning of the nineteenth century plans were developed in England to establish an Astronomical Society (later Royal Astronomical Society, RAS) and to build an observatory on the southern part of the earth. John F.W. Herschel (1792–1871) was, although only 28 years of age, one of the founders of the Astronomical Society. Herschel was the son of the musician-astronomer William Herschel (1738–1822), who came from Hannover in Germany and was the discoverer of the seventh planet Uranus.

The Royal Observatory, Cape of Good Hope was founded in 1820 near Cape Town in South Africa, in order to make star observations in the Southern Hemisphere. The first director was Fearon Fallows (1788–1831), director from 1820 till 1831. Building up an observatory was not easy. A letter to the home front with reports, propositions and requests for extra money took two months to reach England by ship. So an answer took at least four months. The building of the Observatory was a devil of a job. The last stone of the building was laid in 1828, in the presence of the British governor of the Cape Colony Sir Galbraith Lowry Cole. But with all instruments in their place the problems were not over. The main instrument, a copy of the Mural Circle[1] in Greenwich, had small defects, which prevented Fallows to have his name written down in the annals of British astronomy. Thomas J. Henderson (1798–1844) became the second director, from 1831 till 1833. He managed to make the mural circle work properly and with it the positions of thousands of stars of the Southern Hemisphere were measured. Henderson spoke of the Observatory grounds as being a disgusting swamp, quite different from the

---

[1] A mural circle is a telescope mounted on a circular frame, which itself is fixed on a wall exactly along a north-south meridian. It has a scale from 0° till 360°, which allows the angular heights of stars to be measured.

© Springer Nature Switzerland AG 2018
J. Guichelaar, *Willem de Sitter*, Springer Biographies,
https://doi.org/10.1007/978-3-319-98337-0_4

words the young De Sitter in love used to describe the scenic beauty roughly 65 years later. There must have been a fair amount of ants and snakes in those days making life quite unpleasant. The wife of Henderson's assistant William Meadows even versified:

No, it is not like land and it is not like sea

But lest it be asked, where this strange home can be?

'T is as well to confess that it stands on a plain

Over which the eye wanders for beauty in vain.

Henderson had a poor health and after two years he already returned to England in 1833, without having published much work. On his way home at the island of St. Helena he met Manual Johnson (1805–1859). Johnson had been to Fallows in South Africa a few times, as an assistant and to take advice on the building of an observatory on the island. From Johnson Henderson heard that the star α Centauri had a large proper motion. By chance Henderson had done a lot of measurements on α Centauri and back in Scotland he was able to calculate the parallax of the star, leading to a reasonable estimate of its distance. Unfortunately the German Friedrich Bessel and Russian Friedrich Struve got the credits by publishing their later measurements of parallaxes earlier than Henderson.

The Observatory developed further under Henderson's successor Thomas Maclear (1794–1879), director from 1833 till 1870. During the first years of his directorate John Herschel and his wife were at the Cape. The two gentlemen did a lot of measurements together (nebulae, double stars and Halley's Comet in 1836). For all his excellent work Maclear received a knighthood in 1860 and became Sir Thomas Maclear. In 1870, he was then already 76, he wrote his letter of resignation after some pressure from England.

Edward James Stone (1831–1897) was the next director, from 1870 till 1879. He found the Observatory in a bad state of maintenance. But his main aim was to publish at last the star catalogue of the Southern Hemisphere, for which measurements had been made over a large number of years. He was supported in this task by William H. Finlay (1849–1923), his First Assistant since 1873. Finlay discovered the comet 15P/Finlay in 1886, named after him. And he was the first astronomer to observe the Great Comet in the afternoon of 7 September 1882. He stayed at the Observatory till 1898, just long enough for the young De Sitter, who arrived at the Cape in 1897, to meet him. As First Assistant Finlay performed part of the heliometer measurements of the four big moons of Jupiter, then under the director David Gill. These measurements would later be reduced by De Sitter, which led to his dissertation in 1901. At last the catalogue of 12,441 stars of the Southern Hemisphere was ready in 1879 and was published in 1880. After that Stone returned to England (Gill 1913; Warner 1979).

David Gill (1843–1914)

David Gill, descending from a family of clockmakers in Scotland, studied in Aberdeen and followed lectures by James C. Maxwell, famous for his equations describing the electromagnetic field and his proof that Saturn's rings consisted of many small pieces in order to be stable. After his studies Gill learned the profession of clockmaker and joined his father in his business. The professional necessity to measure time exactly brought him in contact with astronomy. It became his great passion and he continued his career in that field. He became an astronomer with James L. Lindsay (1847–1913), a lord who built his own observatory in Dun Echt near Aberdeen. A great success for Gill was his private expedition to Ascension (St. Helena) in 1877. There, with a heliometer borrowed from Lord Lindsay, he was able to determine the distance from the earth to the sun, due to the planet Mars being in opposition. He measured this distance, the so-called Astronomical Unit (AU) with a precision unknown in those days. He received the Gold Medal of the Royal Astronomical Society for his achievement in 1882.

Gill was appointed Astronomer Royal at the Cape of Good Hope as successor of Stone in 1879. He brought a breath of fresh air at the Cape: new methods and techniques, and more instruments. But first the chaos of buildings, papers and instruments had to be brought into some order. In a letter Gill wrote:

Neither soap nor water have been used here over the last nine years. The books are covered with dust and have not been used for 15 or 20 years.

Mrs Stone kept pigs. You can imagine how our garden looks like!! (Warner 1979)

Gill had taken with him Lord Lindsay's heliometer, which he had bought. He used it for parallax measurements. With a widescreen portrait camera he managed to make a beautiful photograph of the Great Comet, earlier observed by Finlay. On that photograph Gill also saw large numbers of stars. It brought him the idea to make star maps with the help of this new method. The labour-intensive determination during the measurements of the positions of stars could be omitted with this new method. Gill's memorandum about the new method to the Paris Academy of Sciences marked the starting point of a new period in astronomy. The first photographs for the star catalogue of the Southern Hemisphere, the Cape Photographic Durchmusterung (CPD) were made in 1885. The German word *Durchmusterung* was used following the Bonner Durchmusterung of 325,000 stars of the Northern Hemisphere published by Argelander, Krüger and Schönfeld between 1852 and 1859.

The Dutch professor of astronomy in Groningen Jacobus Cornelius Kapteyn, who had read about Gill's plans, offered in 1885 to measure the photographic plates, for which an instrument should be developed in collaboration between the two men. The two men became good friends. Gill visited Kapteyn in Groningen on one of his European tours. Together they went to the factory of Repsold in Hamburg, maker of astronomical instruments, where they inspected Gill's new 7-inch heliometer. He said to Kapteyn:

Fig. 4.1 The Great Comet of 1882, first observed by Finlay. Photograph taken by David Gill. The stars produced the idea of measuring star positions with the help of photographs

I would only be half as happy, if you were not jealous. (Kapteyn 1914)

In 1891 Gill, in cooperation with his assistant Finlay, made a large number of observations of the four big moons of Jupiter with the help of his new heliometer. The aim of these measurements was to achieve better ephemerides (lists of positions in the future of celestial bodies).

A heliometer was originally meant to measure the diameter of the sun and its variations. But the mutual (angle) distance between two celestial bodies near to each other in the vision field of a telescope could also be measured accurately with a heliometer. Gill and Finlay measured the mutual distances of the four Jovian satellites. Gill would entrust their measurements to De Sitter seven years later, when he visited the Cape. Gill's later assistant Bryan Cookson (1874–1909) would later add to the bulk of measurements of Jupiter's moons.

De Sitter's mathematical talents made him the right person to reduce these measurements, which had been lying in the archive for seven years. De Sitter's dissertation in 1901 was the result of his reduction work on Gill's Cape measurements.

In 1900 Gill received a knighthood and became Sir David, after having been elected a Fellow of the Royal Society already in 1883. In 1907 Gill retired. Sir David and Lady Gill, who had suffered from a weak health for years, went to live in London. Here Gill kept on entertaining his fellow astronomers, mostly good family friends, and dedicating himself to astronomy.

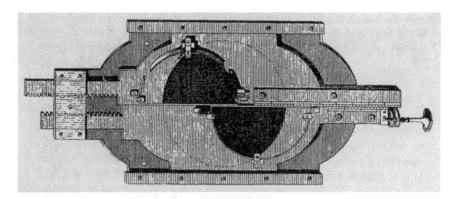

Fig. 4.2 A heliometer made by John Dollond in 1755 [The photograph is taken from E.H. Geyer's article 'The heliometer principle and some modern applications'; *Astrophysics and Space Science*, 110, pp. 183–192 (1985)]. The objective is split in two. The line between the two stars on the celestial globe is taken parallel to the partition line of the objective. Without sliding there are two images. Sliding gives four images in one line. After sliding the two half lenses over a distance d, measured by a micrometer, two of the images are made to overlap. The angle between the two lines of vision to the two stars α is then given by: α = d/f, with f the focal distance of the objective

David Gill Makes De Sitter an Astronomer

We go back to De Sitter in 1896. At the beginning of that year Kapteyn wrote to Gill that he unfortunately had no money to visit the astrographical congress in Paris about the *Carte du Ciel*. But there was a lot to be discussed between the two men. Could Gill and his wife, who did go to Paris, perhaps come to Groningen for a visit of a few days? Kapteyn wrote that in the meantime a pupil of his was busy with the reduction of photographic plates. Gill's answer was positive [RGO, 15/128, 12-4-1896].

In the beginning of October Gill met for the first time in Groningen the then 24 years old student Willem de Sitter, who supported Kapteyn as a calculator. In the introduction to his series of articles about Jupiter's moons in the Cape Annals, Gill wrote his recollections of his first meeting with the young student De Sitter. He saw him at work in the Laboratory of Kapteyn and was struck by the earnestness and capacities of young De Sitter. He offered him a job as calculator at the Cape Observatory, with the important aim to introduce him to practical astronomy.

Much later, in 1911, Gill was busy writing a history of the Cape Observatory. He asked De Sitter to put on paper his recollections to their first meeting in 1896. A part of De Sitter's answer [RGS, DOG/159, 12-8-1911] was:

Fig. 4.3 Sir David Gill, one
of the two teachers of De
Sitter. Photograph from
(Warner 1979)

I was a student in Groningen and worked at Kapteyn's Laboratory, busy with the measurements, discussed in the Groningen Publications 2 and 3. You came to visit Kapteyn.
Although I don't know the precise date, I can recollect the circumstances clearly. On an
afternoon you came in with Kapteyn to have a look at the plates and the measuring
microscope with which I worked. I tried to speak to you a few words in English, but I am
afraid you didn't understand anything of what I tried to say. The following morning, having
breakfast in my room, I received a message from the Laboratory that you wished to speak
with me. I know exactly how you were seated in Kapteyn's room—he was lecturing in the
lecture room—and I stood listening and tried to understand you. At last Mrs Kapteyn came
in and acted as an interpreter. You offered me to go to the Cape as a calculator, and so to
finish my astronomical studies, or rather to start with them, because up to then I had not
occupied myself in particular with astronomy, I had the intention of becoming a mathematician. We agreed that first I would take advice from my parents, that I would finish my
exams, and that after that I would go to the Cape. I arrived at the Cape on 27 August 1897
and returned on 6 December 1899. I came with the intention to perform parallax measurements with the McLean telescope and took with me a complete schedule of activities.
The telescope was not in working order in time, as you know. Besides some measurements
on meteors, comets, occultations[2] etcetera, my observational work at the Cape consisted of
the photometrical work described in the Groningen Publications 12, and the heliometer
observations: parallax of four stars, observations of red stars (appendix to parallax work,

---

[2]During an occultation the light of an astronomical body, for instance a star, is obscured by another
body, such as a planet or a moon. A complete solar eclipse is an occultation of the sun by the
moon.

Cape VIII, 2) and a part of the polar triangulations. The reduction of the parallax observations was entrusted to me in my function of calculator. And after that I started with my
work on Jupiter's satellites under your guidance. PS After I had written this letter, I
managed to find the date of our first meeting in Groningen. It was on 2 October 1896.

Gill, in his book on the history of the Cape Observatory (Gill 1913), gives a
slightly different text.

After Gill's visit to Groningen De Sitter, systematic planner as he was, outlined
his ideas for his visit to the Cape Observatory. They corresponded regularly until
De Sitter's departure [RGO, 15/128, 29-12-1896 until 10-7-1897]. He enjoyed the
prospect of his visit, also by photographs of South Africa Mrs Kapteyn showed
him. Gill wrote that De Sitter had to learn working with all instruments. De Sitter
was a bit disappointed to hear he could not live in the Observatory, but had to rent a
room at 10 min walking distance. In preparation he did spectroscopical work at the
physical laboratory of professor Haga. He wrote to Gill that De Sitter was certainly
able to do more work than that of a simple calculator. Gill was satisfied with De
Sitter's intentions, but wrote that he would have to work extremely hard to finish it
all within one year:

I would be very glad if you could stay longer.

Already in October 1896 Kapteyn wrote to Gill that it slowly dawned on De
Sitter what an invaluable offer he had received. He saw him become ever more
enthusiastic. He added a few months later that De Sitter slowly became reconciled
with the idea of becoming an astronomer. Kapteyn hoped that Gill would make him
a complete astronomer; if not a practical astronomer, than certainly a theoretical
astronomer with a good view about the practical side [RGO, 15/28, 25-10-1896,
26-2-1897]. Kapteyn's insight into De Sitter's possibilities and future was
razor-sharp.

Kapteyn congratulated Gill in May 1897 with his appointment as foreign
member of the Academy of Sciences of The Netherlands. De Sitter had just passed
his final exam[3] *cum laude* (with distinction). Only one in ten manages to reach that,
Kapteyn wrote proudly [RGO, 15/128, 5-5-1897]. He could still do some work after
his own choice. He would like to stay longer than a year at the Cape. Gill had
mentioned a stay until the spring of 1900, but De Sitter considered that a bit too
long and kept his options open. Before deciding he wanted to perform all astronomical exercises first. He would like to be present at the installation of the McLean
telescope and make parallax plates with it. Gill wrote that everything at the Cape
would be in order.

The whole family accompanied De Sitter via Wight to Southampton. From there
De Sitter left on the steamer Hawarden Castle on 7 August 1897. Destination: Cape
Town.

---

[3]Dutch: Doctoraalexamen. This exam made it possible to write a dissertation and get a doctor's
degree.

De Sitter must have been overawed by the Observatory at the Cape, a real observatory with all kinds of astronomical instruments and already a long history. That was something completely different from the office building of Kapteyn's Astronomical Laboratory, to which Kapteyn had moved in the beginning of 1896 (Van der Kruit 2015). After his arrival at the Cape Observatory De Sitter was put to work immediately. Gill wrote at the end of the year to Bryan Cookson, who had asked him how to become a good astronomer and later indeed became one of Gill's assistants:

> I have here a very nice young man. At the table beside me he is busy from 9 until 3 with the reduction of my heliometer measurements of star parallaxes. In the evening he is learning to use the geodetic theodolite[4] and the passage circle.[5] After that the heliometer, followed by the equatorial[6] with the thread micrometer, the photometer[7] and the spectroscope.[8] And before he returns to The Netherlands in about two years, he will have done some independent work by himself. (Forbes 1916)

So Gill made De Sitter do some hard work on all instruments. And De Sitter loved it. At the Cape he became fluent in English. There he met Robert T. A. Innes (1861–1933), who had come to work with Gill in the beginning of 1896. They became friends for life. Together with Innes, who would later become director of the Union Observatory in Johannesburg (from 1903 until 1927), he would later effectuate the collaboration between the observatories of Leiden and Johannesburg.

But in Cape Town De Sitter had not yet lost his interest in mathematics. He wrote a short article about the strophoid, a higher-degree curve, which he sent to his professor of mathematics in Groningen Pieter H. Schoute (1846–1913). Schoute, an expert on curves and multidimensional geometry, wrote back that he enjoyed De Sitter not having forgotten the noble mathematics. Unfortunately Schoute himself had already published on the subject in Crelle's[9] mathematical magazine [ASL, 15-1-1898]. From study notes of De Sitter it shows that he worked on multidimensional geometry in the years before his final exam in 1897, also the Riemannian geometry. Later that would be of benefit to him in his study of the general theory of relativity.

---

[4]With a theodolite angles in the horizontal and in the vertical plane can be measured with great precision.

[5]A passage circle is a meridian or transit circle, measuring the location of a star passing the meridian.

[6]An equatorial is a telescope which is mounted in such a way that one of the two rotational axes is parallel to the axis of Earth. This enables the observer to follow a celestial body by only rotating the telescope at a constant speed (often motorized) around the direction of the Earth's axis.

[7]With a photometer the amount of light received from a celestial body can be measured. John Herschel was one of the first astronomers to use it.

[8]With the help of a spectroscope the light coming for instance from a star can be analysed, measuring the intensities of the different frequencies. From these data many properties of that star can be deduced: chemical composition, temperature, density etcetera.

[9]Crelle was a journal for pure and applied mathematics, named after the German August L. Crelle, its founder in 1826.

Gill proposed to De Sitter in 1898 to use his heliometer observations to calculate the orbital data of Jupiter's moons and the mass of the planet, in order to write his dissertation on the subject. Although Gill had not consulted Kapteyn, the latter did not feel passed over at all, he welcomed the plan.

He is a lucky boy, as he himself is very well aware of.

[RGO, 15/128, xx-1-1898, 3-2-1898]

Gill's offering of his measurements was a stroke of genius. If De Sitter had still remote ideas becoming a mathematician, there was no way back then. He would become an astronomer.

Marriage and Return to The Netherlands

De Sitter had the time of his life at the Cape, certainly after he had met Eleonora (Non) Suermondt. She was born on 16 November in Surabaya on the island of Java in the Dutch East Indies as a daughter of coffee planter Benjamin A. Suermondt (Rotterdam, 1837—Lebak Paré, 1881) and Louise H. van der Wiel (Batavia, 1854). Eleonora's maternal grandparents were Louise A.E.T. Mcgillavry (1826–1874) and Jacobus van der Wiel (1824). Jacobus decided after dismissal from his function as an inspector 1st class of the Water Management in Surakarta on Java to go for a more adventurous job. He started as a private pirate and worked in the Street of Madura, between Java and Madura, under the name of Monseigneur de Mérode. An uncle of the family said never to despair, when your children chose for professions you did not like, because a granddaughter of the pirate married a famous astronomy professor in Leiden (Mac Gillavry 1978). Mother Louise died in a mental home in Buitenzorg on Java. Speaking about her descent was more or less a taboo later in the De Sitter family. Probably, Eleonora had no easy youth. Together with her sister Mary she was sent to the Netherlands shortly after their father's death, where they grew up with two aunts in Rotterdam. After her education she went to work as a teacher/governess in South Africa, where she met De Sitter. Probably it was love at first sight. De Sitter asked her to marry him on the Table Mountain in Cape Town.

It is May. He writes me: when all of us climb the Table Mountain tomorrow, I want to take you to a beautiful spot I know.—There I want to ask you something. Ask something. My God. When we were there on the top, he guided me to it. And life starts.

(De Sitter-Suermondt 1948)

Eleonora's lifelong adoration of De Sitter started in those early days. Her words above are an example of the hagiographic character of her book. De Sitter was also totally in love:

I am completely full of it, like cut glass is full of light.

(De Sitter-Suermondt 1948)

More beautiful an astronomer in the making could not put it to words. It is not known what De Sitter's parents thought about the more or less hasty marriage. But, perhaps at the advice of De Sitter's father, the lawyer, a prenuptial contract was signed, in which was agreed that the intended marriage would be without community of property.[10] In January 1899 Kapteyn wrote to Gill being grateful to him having dissuaded De Sitter from a marriage in his (financial) position. But, taking into account the fact that they were married already, Kapteyn analysed De Sitter's possibilities for the future. From the government Kapteyn had the consent for an assistant's position in his Laboratory, on a monthly salary of 85 lb. He would appoint an assistant on a temporary basis, keeping the post available for De Sitter on 1 January 1900. The salary was hardly sufficient for an unmarried assistant. But Kapteyn looked at it from the bright side, expecting De Sitter's father to support him while building his scientific career. And moreover, Kapteyn saw possibilities for some additional salary. De Sitter certainly wished to work with Kapteyn. According to Kapteyn De Sitter could show in Groningen what his qualities were, preparing himself for a vacant professorship in astronomy or mathematics, as long as that vacancy would not be too far in the future [RGO, 15/128, 3-1-1899]. Gill and Kapteyn completely agreed on the talents of De Sitter. Thus Kapteyn was already thinking about a professorship for De Sitter. We can read his thoughts: in Groningen Kapteyn would stay until 1921, in Utrecht Nijland had been appointed only recently, in 1898. Only in Leiden Van de Sande Bakhuyzen would retire in nine years. That had to be the aim. So there were nine years to write a dissertation and to build up a good list of solid publications. A professorship in mathematics was another possibility, but that had become less obvious in the meantime. About the marriage the two gentlemen did not worry anymore. Perhaps they could appreciate the self-willed couple. Gill wrote to Kapteyn:

> De Sitter is deliriously happy; he is however working away quite steadily and not allowing
> Cupid to upset his work. (De Sitter-Suermondt 1948)

At the end of 1899 a son was born in South Africa, Lamoraal Ulbo, and with melancholy they returned to The Netherlands just before the turn of the century, with their baby and a substantial pile of heliometer measurements. The return trip took longer than planned, because the engines of the steamer were dirty. With the baby in arms or in a basket travelling by train from Southampton through England and by taxi through the fog and cold of London must not have been easy. At the beginning they stayed at De Sitter's parents in Arnhem. Father, mother, brother and sister fought for the honour to look after the baby. In a letter to Gill De Sitter wrote that he could hardly imagine not sitting at the table with the wooden chair on the next working day. But wherever he would work in the future, that table and chair

---

[10]Family archive in the possession of De Sitter's granddaughter Tjada van den Eelaart-de Sitter, Arnhem.

would always have to do with it [RGO, 15/128, 4-1-1900]. On Monday 22 January the De Sitters went to Groningen. First they stayed with the hospitable Kapteyns, later with an aunt of De Sitter, till their new home was ready.

Fig. 4.4 The young married couple Willem and Eleonara. Photograph from the family archive

Writing a Dissertation

But then work on De Sitter's dissertation had to start seriously. Kapteyn allowed him ample time to get to work on the reduction of the measurements of Jupiter's moons. So for Kapteyn's own work De Sitter was not very useful, but he still was of the opinion that he could not wish for a better assistant [RGO, 15/128, 9-3-1900]. In the end, after his dissertation, De Sitter published a number of excellent articles within the research field of Kapteyn. De Sitter had received a lot of material from Gill: the heliometer measurements Gill had made with Finlay, his assistant until 1898. So De Sitter and Finlay worked together at the Cape for one year. Gill had built up an increasing trust in De Sitter, on which basis he had entrusted to him this important set of measurements. De Sitter would effectuate the research with the utmost precision. Then he knew that a thorough dissertation would be of great interest to the rest of his career. He hoped to be able to finish his dissertation within a year, but he had still to receive material from South Africa. Corresponding and sending further information from the Cape took much time and influenced the planning of the doctoral degree considerably. De Sitter wished to see and check everything himself. Results of calculators at the Cape should not be his basis. If anything should appear to be inaccurate, he would be blamed. Sometimes a photographic plate was lost, which had then to be done again. The distance between Groningen and Cape Town was a real problem. After receiving a letter from the English astronomer Ralph A. Sampson (1866–1939), who also worked on Jupiter's moons, De Sitter became nervous [DUL, 4-11-1900]. De Sitter had written to him on the advice of Gill. On the whole De Sitter's results did not correspond to Sampson's own, and he made clear that he hardly trusted De Sitter's results. The further contacts between Sampson and De Sitter will be discussed in the next chapter. De Sitter wished to run over the calculations again. He had to know mistakes before publication. It would be horrible if someone else would find mistakes, or worse that mistakes would not be found at all. This characterized De Sitter as a real scientist: a not-found error is worse than the shame about an error found by someone else. He indeed came across some errors, for instance one of 10 min in the Greenwich time, the correction of which would cost him much time. But the dissertation had to be ready before the summer of 1901. He wished to start the academic year 1901–1902 with lectures as a private lecturer.[11] He insisted a number of times with Gill on receiving calculations that were done by calculators at the Cape. His dissertation would be the basis of his work in the coming ten years. Gill would certainly endorse De Sitter's responsibility to verify everything, he wrote. And with a bit of drama:

> My whole career as an astronomer depends on this work.

Gill tried to reassure De Sitter. The small errors would probably be of little influence on the values [RGO, 15/130, 20-8, 8-9, 28-11, 6-12, 19-12, 1900].

---

[11]A private lecturer (Dutch: privaatdocent) was an unsalaried university lecturer.

In Groningen a Dutch marriage certificate was drawn up on 8 August 1900, after the marriage in South Africa had been legalized by the Dutch Ministry of the Interior on 8 June. On 28 September 1900 the De Sitters had another child: a daughter, Theodora.[12] But then fate struck suddenly. Their son became fatally ill and died on 10 January 1901. On the same day De Sitter wrote to Gill:

> Dear Sir David, I was very glad to receive your letter with the details of a number of substitutions. I am not in the mood for writing at the moment. Our dearest boy died this morning after an illness of nearly five weeks. Give my best wishes to all. Always yours, W. de Sitter.

The baby was only fourteen months. A couple of weeks later De Sitter wrote again, this time with a nearly religious undertone:

> But we have to accept our fate, and try to think that it is all for our own good, although we cannot understand it.

On the envelope he stuck a number of special stamps for the collection of his friend Innes. Gill answered that gloom hung over the Observatory and that he wished that God might help and console them [RGO, 15/130, 10-1, 29-1, 30-1, 1901]. The times must have been hard for the young couple, this unfathomable disaster after a few years of complete happiness. In retrospect we may perhaps state that these years of happiness and sorrow made their strong marriage. Non and Willem, as someone later said, were able to make the happiness in their further life. De Sitter looked for comfort during these days by working hard at recalculations and checking. Already on 7 February he wrote Gill that he had finished. All bigger and minor errors had done little harm. Besides he asked a number of detailed questions about the English language. For example:

> How do you mention a group like a_1, a_2 and a_3? Do you say the a's or the a?

So with or without the plural s. Gill had read the texts with great pleasure and he had taken notice of a number of remarkable conclusions of De Sitter concerning further research. After a letter from Gill De Sitter could write to Gill's satisfaction that now, after all adjustments, there was also a reasonable agreement with Sampson's results [RGO, 15/130, 7-2, 14-3, 17-5, 25-5, 1901]. Nowhere in the correspondence between Gill and De Sitter concerning the dissertation Kapteyn appears on the scene. Because Kapteyn and De Sitter worked together in Groningen, there is no correspondence between them in this period. Still it seems that Kapteyn did not play an important role in the realization of De Sitter's dissertation. That is understandable for a number of reasons. First the dissertation was completely based on Gill's own observations. Moreover, it was Gill's plan to make the heliometer observations of Jupiter's moon the subject of the dissertation. It is true that Kapteyn had immediately consented to the chosen subject, on the basis of his high esteem for and complete trust in Gill, but it was not a subject in his own

---

[12]*Municipal register of Groningen*; Archives of the province of Groningen (Dutch: Groninger Archieven) in Groningen.

field of research (parallaxes, proper motions, dynamics of large groups of stars). Another reason was perhaps that Kapteyn was in the middle of his calculations and analyses regarding the so-called star streams. In retrospect Kapteyn must have been thinking about a subject for De Sitter's dissertation in an earlier stage, a subject in the field of the analysis of the structure of the universe. He could certainly have made good use of De Sitter's mathematical talents. Probably Gill was simply swifter than Kapteyn. Perhaps he was afraid that his beautiful heliometer measurements would in the end not be reduced. Anyway, the subject was Jupiter's moons.

The day of the doctoral degree was set on 17 May 1901. De Sitter took his doctor's degree *cum laude*. He even received two degrees. The first as doctor in mathematics and astronomy, in the afternoon at 14:30 h, on his dissertation *Discussion of Heliometer-Observations of Jupiter's satellites, Made by Sir David Gill K.C.B. and by W.H. Finlay M.A.* (De Sitter 1901). An hour later, at 15:30 h, he got the second title as doctor in mathematics and physics on the basis of 26 propositions. The number of printed copies was 300. The dissertation was circulated widely in the national and international astronomical world.[13] On 5 June Gill received four copies: one for himself, one for Innes, one for his chief-assistant Hough[14] and one for the library of the Observatory. He congratulated De Sitter with his *magnum opus* and his *cum laude*. Gill congratulated himself too with his heliometer observations, which were obviously satisfactory in every respect. Gill was very satisfied with the analysis of the results. He immediately wished to do further observations with his in the meantime overhauled heliometer. In particular he was enthusiastic about the fact that De Sitter had probably found the so-called Laplacian libration [RGO, 15/130, 5-6, 10-6, 1901]. This phenomenon is based on the special ratio, commensurability, of the orbital periods of the inner three large satellites of Jupiter, Io, Europa and Ganymedes: 1:2:4.[15] De Sitter would later use this specific ratio for his new analytical theory of the four large moons, including Callisto.

After a holiday in June with his parents in Arnhem, where De Sitter and his wife walked a lot on the moors and in the woods, De Sitter started to prepare his lecture series as a private lecturer. He wrote to Gill to be glad with his positive comments and, as a young doctor, he ventured a small joke by remarking that he found the typewriter that Gill had used for the first time not an improvement compared with his handwriting, which he liked more [RGO, 15/130, 10-6, 11-7, 1901]. In the Gill Archives of the Royal Greenwich Observatory in the University Library in Cambridge not only the letters received by Gill are found, but also copies of the letters sent by him. They are ink copies on very thin paper, on which the ink is

---

[13] ASL, Circulation list, letter of E.B. ter Horst, 17-5-1901.

[14] Sidney S. Hough (1879–1923) became first assistant of Gill in 1898 and director of the Cape Observatory after Gill's retirement in 1907.

[15] See Chap. 5.

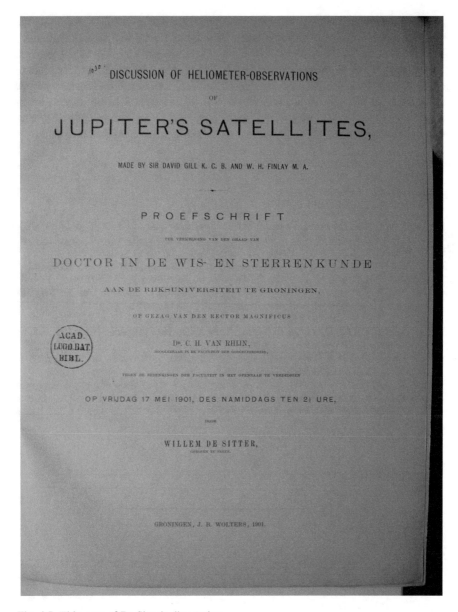

Fig. 4.5 Title page of De Sitter's dissertation

flowed out. The paper is so dry and brittle that the merest touch makes a piece of the paper fall out.

Now the road to a career in astronomy lay open for De Sitter. On 1 October 1901 he gave his public lecture as a private lecturer at the Groningen University. He had only one student, while he had hoped for three [RGO, 15/130, 2-10-1901]. Few astronomers have had two teachers of such outstanding quality. Kapteyn, the absolute number one of Dutch astronomy, and one of the three British Astronomers Royal in those years: Gill of the Cape Observatory, Christie of Greenwich Observatory[16] and Copeland of Scotland.[17] The start was flying. But in the years to come there would be a lot of doctoring at the results of the dissertation. And that would not go off without a hitch. Moreover, there had still to be done a lot of publishing to reach the professorial level. And besides in The Netherlands, with three observatories and three professors of astronomy, the market was small.

References

De Sitter, W. (1901). *Discussion of Heliometer-Observations of Jupiter's satellites, Made by Sir David Gill K.C.B. and by W.H. Finlay M.A.*, J.B. Wolters, Groningen, 1901.
De Sitter-Suermondt, E. (1948). *Een menschenleven*. See References of chapter 3.
Forbes, G. (1916). *David Gill, Man and Astronomer, Memories of Sir David Gill, K.C.B., H.M. Astronomer (1879–1907) at the Cape of Good Hope*, John Murray, London, 1916.
Gill, D. (1913). *A History of The Royal Observatory, Cape of Good Hope*, Published by His Majesty's Stationery Office, London, Printed by Neill & Co, Edinburgh, 1913.
Kapteyn, J.C. (1914). 'Sir David Gill'; *The Astrophysical Journal*, X, L, 2, p. 161, 1914.
Mac Gillavry, A. (1978). *For I knew your father* (Dutch: *Want ik heb uw vader gekend*), p. 172, De Kern, Bussum, 1978.
Van der Kruit, P.C. (2015). *Cornelius Kapteyn, Born Investigator of the Heavens*, Springer, Cham Heidelberg New York Dordrecht London, 2015.
Warner, B. (1979). *Astronomers at the Royal Observatory Cape of Good Hope, A history with emphasis on the nineteenth century*, A.A. Balkema, Cape Town and Rotterdam, 1979. Much of the information about the history of the Royal Observatory Cape of Good Hope can be found in this book.

---

[16]William H.M. Christie (1845–1922) was Astronomer Royal at Greenwich Observatory from 1881 until 1910.

[17]Ralph Copeland (1837–1905) was Astronomer Royal for Scotland from 1889 until 1905.

Chapter 5
The Galilean Moons of Jupiter

> *For me Scotland is Sir David Gill; he is the first Scot I have known. I knew him so well. He was simple as a child and big as a highland ben.*
>
> Willem de Sitter (De Sitter-Suermondt 1922)

The Discovery of the Moons by Galileo

There are reports earlier than the year 1600, even Chinese reports from before the Christian era, of observations with the naked eye of the two largest moons of Jupiter: Ganymedes and Callisto. But only with the arrival of the telescope reliable observations were possible. Hans Lippershey (c 1570–1619) and Sacharias Jansen (c 1585–1632), two Dutch lens grinders from Middelburg in the province of Zeeland, made the first telescopes. They were telescopes with a positive objective (convex lens) and a negative eyepiece (concave lens), later called a Dutch telescope. Lippershey received the credits for the invention, because his telescope is the first which was well documented. He showed his telescope to Prince Maurits, stadtholder of the United Provinces of the Netherlands from 1585 until 1625 and son of William of Orange, in the beginning of October 1608. He thought that the Prince would be well aware of the military value of it. He applied for a patent on his invention. From the States General he received only a small grant for further development (King 2003). Prince Maurits and a number of European leaders, met at that time to negotiate the Twelve-Year Truce[1] (1609–1621) during the Eighty Years' War[2] against Spain. Probably through that channel the information on the first telescopes reached Galileo (1564–1642) in Padova in Italy. On the basis of this information Galileo made his own design and made his own telescopes in 1609.

---

[1]Dutch: Twaalfjarig Bestand.
[2]Dutch: Tachtigjarige Oorlog.

© Springer Nature Switzerland AG 2018
J. Guichelaar, *Willem de Sitter*, Springer Biographies,
https://doi.org/10.1007/978-3-319-98337-0_5

He directed his telescope to have a close look at the heavenly phenomena. In a short time he discovered a number of new and spectacular facts: mountains on the moon, sun spots, the phases of Venus, two *ears* on the sides of Saturn (Christiaan Huygens discovered later they were rings), and the four largest satellites of Jupiter. Galileo saw these with a telescope with a magnification of 30. On 7 January 1610 he saw the first three: bright small *stars* at the sides of Jupiter, two to the east and one to the west. On 8 January he saw all three of them to the west. On the 10th there were suddenly four satellites. The moons and Jupiter were nearly exactly in line. In the course of those weeks Galileo realized that these small *stars* were satellites of Jupiter. Already in March his book *Sidereus Nuncius* or Stars Messenger appeared, in which he published his discoveries (Mudry 2005). The German astronomer Simon Mayr (1573–1625) later stated to have observed the Jupiter moons a few days before Galileo. It seems he did it a few days later. But Mayr did propose names from the Greek mythology for the moons: Io, Europa, Ganymedes and Callisto, four lovers of the Olympian god Zeus (Latin: Jupiter). The first quantitative measurements on Jupiter's satellites were eclipse measurements: the times at which a satellite disappeared behind the planet Jupiter and appeared again. From these measurements the first data concerning the Jupiter system could be deduced. The Danish astronomer Ole Rømer (1644–1710) could determine the speed of light with the help of eclipse measurements. Because the eclipses should in principle appear with equal intervals, Rømer was surprised about his observation that, when the earth and Jupiter were on the same side of the sun the eclipses came *too early*, and when the earth and Jupiter were on different sides of the sun, they came *too late*. He supposed that a finite speed of light, instead of an infinite speed, was the explanation. The different time intervals between the eclipses and the distance from the earth to the sun, which was roughly known, led to an estimated speed of light of 225,000 km/s. It took some time before his supposition was accepted.

The Jovian system with its four big satellites resembled the solar system, for instance the sun with the four inner planets, or the sun with the four outer planets. Therefore the data from the Jovian system could be used to gain a better insight in the development of the whole solar system. It was a miniature solar system, in which the movements were a lot faster: orbital periods from 2 till 17 days for Jupiter's satellites, as opposed to 0.24 till 165 years for the planets. From measurements of the Jovian system over a relatively short period conclusions could be drawn about the properties of the solar system over a much longer period, for instance about the question of the stability of the planetary orbits.

Perturbation Theory

The simplest problem in celestial mechanics is the two-body problem. The general solution is that both bodies each describe a fixed ellipse with their common centre of gravity as one of the foci in each ellipse. If one of the two masses is much smaller than the other one, the large mass may be placed in one of the foci of the ellipse described by the small mass. For the three-body problem a general solution cannot be given. A system of three bodies leads quickly to very complex movements. Only in special cases, in which one or two masses are very small compared to the third one, solutions of the three-body problem are possible.

Three-body problem and chaotic movement
In the figure we see a planet P in a circular orbit around the sun S. A comet comes in its elongated elliptical orbit near the planet P. By the gravitation of the planet the comet is 'switched' into a completely different orbit. The comet switches from orbit A into orbit B. Such a near approach may occur again after a number of revolutions, after which the comet proceeds again in a completely different orbit. A minutely different passage along the planet will cause a completely different orbit. Although the system is completely deterministic, the exact orbit of the comet will be unpredictable in the long run.

The problems in analysing the orbits of still more bodies are of course correspondingly greater.

One manner to make predictions of the orbits is, starting with all places **x** and velocities **v** (vectors, written in bold type, while they are three-dimensional) of the masses m in the system known at a certain time t, to calculate all places and velocities a very small time Δt later. The place **x′** and velocity **v′** (Δt later) of m can be found from **x**, **v** and the force **F** by simple mechanical calculations. The working

Fig. 5.1 Three-body problem and chaotic movement

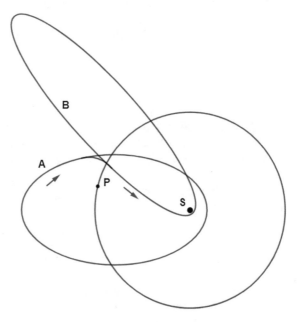

force **F** is the vectorial sum of all the gravitational forces by all other masses on m. Because **F** has not necessarily the same direction as **v**, the direction **v′** may differ from that of **v**. From there we proceed again Δt in time to find **x″**, **v″**, after first having calculated the new force **F′**. And so on. This theoretical correct method leads to extremely large numbers of calculations. Before the computer era this method was not a real option.

A different method, which may be applied, is the one using global calculations followed by the effect of small disturbances or perturbations. A situation which is often the case in our solar system is the one in which the masses differ enormously from one another. In the Jovian system, in view of the large distances and/or small masses, we may neglect the influence of the other planets, of the planetoids and comets. The satellites of Jupiter in their turn have no influence on Jupiter's orbit round the sun. The Jovian system (including the satellites) moves thus in a normal Keplerian[3] ellipse round the sun. In the first approximation each of the four moons of Jupiter describes an ellipse round the planet. The movement of each moon is influenced by several sources. First by the three other moons. Then by the sun, which, it is true, is at an enormous distance, but does have an enormous mass. And also by the fact that Jupiter is not a perfect sphere. By its fast revolution speed the planet is rather oblate. A perfect sphere (with a homogeneous mass distribution) causes forces as if all mass is in the centre. That is not true if the planet is oblate.

Still there is a special circumstance which helps us enormously with the solution of the Jovian system. The disturbing forces may not as said be neglected, but they are still so small that the orbits of the moons are nearly exact ellipses. As a good first approximation of the real orbits we may therefore take ellipses. We now are in a position to solve the problem by successive approximations. We can calculate for each orbital element of a moon (for instance its semi large axis, its eccentricity, the direction of its orbital plane) its change per unit of time. The basic form of the orbit stays an ellipse, but one that changes slowly. But the changes in time of these orbital elements are given by complex equations, which can only be solved step by step in successive approximations. A next order in the calculations gives each time a better approximation.

Orbital elements of planets or satellites

In order to determine the elliptic orbit of a planet relative to the ecliptic (the plane through the orbit of the earth around the sun) six elements are necessary.

Two elements describe the form of the orbital ellipse:

1. The semi-major axis a. We have a = ½AP, with P the periapsis (for a planet perihelium), the shortest distance from the planet to the sun, and A the apoapsis (aphelium), the longest distance from the planet to the sun.

---

[3]Johannes Kepler (1571–1630) was the first one to deduce from the known observations the elliptical orbits of the planets round the sun, his first law.

2. The eccentricity e. We have: $e = \sqrt{(1 - b^2/a^2)}$, with a the semi-major axis and b the semi-minor axis of the ellipse.

Two elements describe the position of the orbital plane of the ellipse relative to the ecliptic.

3. The inclination i. This is the angle between the ecliptic and the orbital plane of the ellipse. In the figure the plane of the drawn angle i is perpendicular to SN. In the ascending node N the planet passes from under to above the ecliptic.

4. The longitude of the ascending node Ω. This the angle in the eliptic from the direction of the vernal point Υ to SN. The direction of the vernal point in the ecliptic is fixed by the orbit of the earth, i.e. when the equatorial plane of the earth passes through the sun. This happens twice a year, at the equinoxes, when day and night have roughly the same length. The vernal equinox happens between 19 and 22 March.

There is still a freedom of rotating the ellipse in the orbital plane. We must still fix the direction of the semi-major axis of the ellipse.

5. The argument of the ascending node ω. This is the angle between the direction of the ascending node and the semi-major axis of the ellipse.

With these five quantities the orbit is completely fixed. The place of the planet in its orbit may now be fixed by a sixth quantity.

6. The True anomaly v. This is the angle from the periapsis to the radius of the planet SP.

For a satellite relative to a reference plain (usually the plane through the planet's equator) comparable elements can be defined.

Pierre-Simon Laplace was the first to develop a mathematical theory to calculate these perturbations. On the basis of his general theory he also developed the first theory for the Jovian system, by using a number of known data in his general theory. If the calculation of the first perturbation did not agree with the observations, it was possible to calculate the second order perturbation, et cetera. That perturbation calculations are extremely complex, can be guessed. The big advantage of a good theory was that it yielded the formulae for the changes with time of the elements. These formulae contained the time t and the times in the future had only to be inserted to calculate the values of the elements in the future. On the basis of these formulae it was then possible to make tables giving for each time in the coming period the place in the sky of each moon in relation to Jupiter. These tables are called ephemerides. Because the perturbations were only calculated up to a certain order, such ephemerides would only give the correct values for a limited

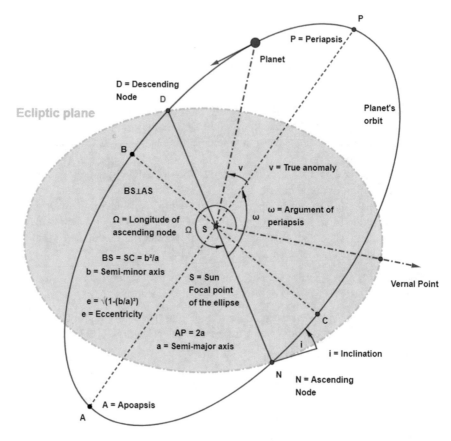

Fig. 5.2 Element of a planet or satellite

period of time. After a period of a number of decades new ephemerides had to be calculated.

Pierre-Simon marquis de Laplace (1749–1827)
The French mathematician Laplace was of humble origin. After his studies in Caen in Normandy he became a professor at the Military Academy in Paris. Soon he became a member of the Academy of Sciences. He developed the theory saying that the planetary system originated from a rotating nebula. Besides he published about the stability of the planetary system. In his later life he was politically active a number of times and acquired then his noble title. Laplace was the first to write a theoretical treatise on perturbation theory in celestial mechanics. He published his *Traité de Mécanique Céleste* in five parts from 1800.

Fig. 5.3 Pierre Simon de Laplace

De Sitter's Dissertation

In the introduction in his dissertation De Sitter gave an overview of the developments until 1900 of the data of the four Jovian moons. The first mathematical theory of the mutual disturbances of Jupiter's moons was, as mentioned, developed by Laplace.

The astronomer Alexis Bouvard (1767–1843), director of the Paris Observatory, calculated a number of quantities in this theory, but many were still unspecified. Jean-Baptiste Delambre (1749–1822), mathematician and astronomer, determined the tables of these quantities in 1792 for the first time on the basis of 2000 eclipses of Jupiter's satellites. Thereupon Laplace made a number of improvements, after which Marie-Charles Damoiseau (1768–1846) published ephemerides in 1836, valid until 1880, which predicted the most relevant events of the moons, like eclipses. Thus the first relevant results were mainly obtained by French scientists. But already in 1842 the German Friedrich Wilhelm Bessel (1784–1846) wrote that better tables were necessary. He said in his *Astronomische Untersuchungen*, published in 1842, that it would be wise to abandon Damoiseau's results and to do new research of the elements. The British Charles Todd (1826–1910) and John C. Adams (1819–1892) each made adaptations to the tables of Damoiseau, making them valid until the beginning of the twentieth century. But there were considerable differences between the predictions and the observations. The most severe flaw was the absence of an underlying consistent theory. A big problem was that one scientist used a number of more or less known values of relevant quantities in order to calculate a number of other quantities, while another scientist made different choices. In that manner they created uncertainties about the values and their accuracy. For example: Adams found for the mass of the fourth satellite roughly half the value obtained by Laplace, and for the mass of the first satellite twice the value obtained by Damoiseau. There were too big differences between theory, calculations and observations to leave the case as it was. Still, it must not be concluded that up to that time poor work had been delivered. Laplace's perturbation theory was a masterly piece of mathematical work, which laid the basis for the ensuing astronomical perturbation theories. And, to achieve later elaborations thereof, we must take into account the unimaginable amount of calculations that had to be performed, on the basis of the observations. And these observations had not yet the accuracy that was achieved later.

In 1880 Cyrille Joseph Souillart (1828–1898) published in the *Memoirs of the Royal Astronomical Society* in London a theory of Jupiter's moons that was in the opinion of De Sitter in many respects more complete than that of Laplace: *Théorie analytique des Satellites de Jupiter*. Souillart's theory would form the basis of all work on Jupiter's satellites in the following decades.

Cyrille Joseph Souillart (1828–1898)
The French Souillart was of humble origin, but after his studies he became a professor in Saint-Omer, Nancy and Lille. His dissertation was about the theory of Jupiter's moons and he worked on them over a period of thirty years. De Sitter did the same half a century later. In 1865 he published *Essai sur la Théorie Analytique des Satellites de Jupiter*, followed by extensive publications in 1880 and 1889. The French mathematician and astronomer François F. Tisserand (1845–1896), who published a four-part standard work on celestial mechanics, wrote about Souillart: The theory of the Jupiter moons by Laplace is unsurpassed. Still some calculations have not been carried through far enough to provide the tables deduced from them the necessary accuracy. It is Souillart's merit to have added these additions. In the first of his two main publications Souillart presented an algebraical theory. In the second one he substituted numerical values for the symbols. But he left the production of ephemerides on the basis of his results to other astronomers.[4]

In the field of observations Gill and his assistant Finlay had made big steps forward. From 17 August until 8 December, 45 evenings/nights in total, they measured the angular distances between each of the six pairs from the four moons, and also their place in relation to a number of neighbouring fixed stars. De Sitter wrote that such a measurement, between two points, was much more accurate than the angular distances between the moons and the rim of the planet disc, as measured by Bessel and later Wilhelm Schur (1846–1901) with their heliometers. The rim was after all not completely sharp, and the observations as a result inaccurate. Only in 1898 Gill offered his up to that time still unprocessed heliometer measurements to De Sitter. The last was allowed to use the calculators working at the Cape and later those of Kapteyn's Laboratory for his computations on the analytical theory

---

[4]'Obituary Notice'; *MNRAS*, 59, 5, pp. 233–234, 1899.

and the subsequent extensive numerical calculations. De Sitter was very grateful to Gill and Kapteyn. He concluded his Introduction with:

> It will be mainly due to Kapteyn and Gill, if ever it will be given to me during my career to do something for the advancement of the science they taught me to love.

Chapter 1 contained the results of the observations and the description of the technical details. Chapters 2 and 3 contained the description of the calculations and their results. He mentioned that he had made use of the *Data to calculate the positions of Jupiter's satellites* of the German astronomer Albert Marth (1828–1897). In Chap. 4 a long analysis of the uncertainties in the measurements and the results of the calculations followed, compared to those of old visual eclipse observations and new photographic ones (where angles between celestial bodies were found by measuring distances on a photographic plate). The Irish astronomer Ralph A. Sampson had already worked on photographic eclipse measurements, that had been made from 1878 at Harvard College Observatory near Boston (USA), and he had already communicated a number of his results to De Sitter. De Sitter came to the conclusion that even more recent eclipse observations contained appreciable errors, that the heliometer measurements were much better and that from photometric observations in the end even more accurate results could be expected.

Ralph Allen Sampson (1866–1939)
Sampson studied at St. John's College in Cambridge. After appointments in London and Cambridge he became a professor of mathematics at Durham College in Newcastle-upon-Tyne in 1893. In 1895 he was appointed professor of mathematics in Durham, where he also became director of Durham Observatory. He did fundamental research on the satellites of Jupiter. Already in 1895 he published about Jupiter in *The Observatory* for the first time, and in 1910 he published the *Tables of the Four Great Satellites of Jupiter*, giving their positions from 1850 until 2000. The Royal Astronomical Society (RAS) published his Theory of the Four Great Satellites of Jupiter in 1921. For all his work he received the Gold Medal of the RAS in 1928. He was president of the RAS from 1915 until 1917. From 1910 until 1937 he was a professor at the University of Edinburgh. He also became Astronomer Royal for Scotland. Once, when by accident he was locked in the clocks' cellar, he was proud to be able to have freed himself within an hour by opening the lock with the help of his pocket knife.[5]

De Sitter considered the precision of his results not yet sufficient to determine all elements of the satellites of Jupiter accurately. He made an analysis of the supplementary data needed to calculate these elements with such accuracy that on the basis of those new ephemerides could be made. In the end he reached a number of

---

[5]*The Observatory*, 63, p. 105, 1940.

recommendations for further research: recalculation of old measurements and implementation of a series of new measurements. He also gave the times of the coming three oppositions of Jupiter as appropriate opportunities to do new observations: 1902, 1903 and 1904.[6] So De Sitter did not stop at the reduction and more accurate determination of a number of elements of Jupiter's moons from the heliometer measurements. With this analysis he set himself the task to continue the inquiries into the moons of Jupiter. They had him under their spell. It would, with a few doubts over the years, keep him busy, figuratively speaking every day over a period of thirty years. He would spend more time and energy in it than in his later work on the general theory of relativity, on which he worked intensely from 1915 until 1918, and later from 1929 until 1933.

The Sampson Case

De Sitter's joyful doctoral degree *cum laude* was only a few months back before Sampson, with whom he had corresponded while working on his dissertation, threw a spanner in his works with a critical commentary in *The Observatory* of July (Sampson 1901a). He considered De Sitter's discussions admirable in many respects, and his calculations formidable. Therefore it was with great regret he had to point at some faults, which in his opinion were very serious. The first point dealt with the use of the four distances of the moons to Jupiter. De Sitter had eliminated these four unknowns on the basis of data from the tables of Marth. But just that was not allowed according to Sampson. This assumption of De Sitter Sampson called ironically *an act of faith*, which he was not able to follow. The second point concerned the mathematical relation used by De Sitter between these distances and the mass of Jupiter. According to Sampson this relation had to be replaced by a more complex one. After the correspondence preceding the publication of his dissertation De Sitter thought that the differences with Sampson were for the most part cleared away. The criticism made him furious. He wrote to Gill from his holiday address in the country, where the family stayed for the health of their daughter, if he ever had heard such impudence before?

> I do not believe that so many blunders were ever written in such a short text.
>
> Of course it is all nonsense.

De Sitter could not understand it. Sampson had tried to find the tiniest error. He had already sent an answer to *The Observatory* (De Sitter 1901). After having

---

[6]In an opposition two celestial bodies are seen from a third one at opposite points on the celestial sphere. For Jupiter it means that the earth is in between Jupiter and the sun, which means that the distance from the earth to Jupiter is a minimum. As a result of the large orbital period of Jupiter round the sun (12 years) there are oppositions with just over a year in between.

thanked Sampson in the opening lines of his letter for the interest in his work, De Sitter continued with undisguised sarcasm.

> I am glad that professor Sampson did not find more serious objections to my work than these two, which both are no doubt a result of a misunderstanding on his side.

Little explanation was necessary. He wrote with reference to his dissertation that he had carefully examined Marth's data concerning the distances and compared them to Souillart's theory. He had mentioned the errors, but after that he had not felt obliged on each use of a value of Marth to mention it was explicitly correct. He defended his calculations with vigour. And he did not have to resort to the faith, because the distances of Marth could be verified with a very simple calculation. De Sitter also disposed of the second point of criticism, the too simple connection between the mass of Jupiter and the distances of the moons.

For an astronomer, who had only very recently written his dissertation, in fact a novice in science, his tone towards Sampson, six years his senior and having been a professor since 1893, was a bit impertinent. As would appear later, a more careful answer would have been wiser. The old Gill must have shaken his wise head on reading De Sitter's furious letter about Sampson's *blunders* and *nonsense*. He must have waited anxiously for the next edition of *The Observatory* to read De Sitter's rashly written answer. Sampson let not pass the snub by De Sitter and he made clear again his points in October (Sampson 1901b). The readers would then be able to decide without difficulty if his objections are trivial or not, and on whose plate the errors really lied. And Sampson again analysed De Sitter's calculations scrupulously. He finished with an analogy. Many good ships had been wrecked (during the calculations of the corrections on the orbits of Jupiter's moons). And De Sitter, the most recent skipper, had not been able to avoid the rocks. He added that each newcomer could be asked with emphasis if he had done what he claimed and if he had taken with him all his cargo safely.

> And, at least at one point, I say again that Mr. De Sitter has failed.

It turned out to be that Sampson was partly right. De Sitter wrote to Gill that Marth had omitted perturbations of the first order. During his checks of the distances De Sitter did not have part II of Souillart's work at his disposal. If so, he would very probably have noticed it. With that knowledge he could easily calculate the effect on the mass of Jupiter. He would write to Sampson. Regarding the second point, Sampson was of course completely wrong [RGO, 15/130, 17-10, 13-11, 5-12, 1901]. There followed another two comments by De Sitter in *The Observatory*. Sampson finished his contributions to the controversy with a few neutral remarks: De Sitter had acknowledged that Marth had used an erroneous formula for the calculation of the average distances, and that he in consequence had recalculated his final results.

A postscript of the editors followed, thanking the gentlemen for their discussion, which had been instructive for their readers. But they did hope that De Sitter would not deem it necessary to write another addition. De Sitter could take that to heart, although there appeared a last short comment from his hand, with the statement that the truth in this discussion did not come nearer and that Sampson only came with statements and not with proofs (De Sitter 1902).

Sampson only wished to win his case, without keeping an eye on the truth, De Sitter wrote to Gill. He would not write another answer, he was disgusted by it. They did analyse once more the in their eyes incomprehensible behaviour of Sampson. But after all, De Sitter had better asked advice from Gill, or Kapteyn, before his first precipitate answer to Sampson.

But for De Sitter this struggle was by no means a reason to give up working on Jupiter's moons. On the contrary, he sunk his teeth into it, but chose to embark on a sensible course. He decided to reduce the photographs being made at the Cape. He would concentrate on the inclinations and nodes of the orbits of the moons. Sampson was mainly busy with the elaboration of eclipse observations, made at Harvard College Observatory, which would lead to a number of other data, the masses and at last the ephemerides. De Sitter thought that Sampson was very well capable to do that work properly. Only when Sampson would not do the job, De Sitter would perhaps take it up himself [RGO, 15/130, 12-2, 5-3, 27-3, 29-6-1902]. Kapteyn and Gill did worry about the row, but they supported De Sitter. Kapteyn had heard in England that both Sampson and Hough, first assistant at the Cape, had offered their help to De Sitter, but that he had refused it [RGO, 15/131, 11-3-1902 Kapteyn to Gill]. The reasons of the heated letters to *The Observatory* may be partly understood from the preparatory phase until the publication of the dissertation. Gill had advised De Sitter ample time before his doctoral degree to write to Sampson. He had even sent an introductory letter to be used by De Sitter. Probably Gill informed Samson of the coming letter from De Sitter. Perhaps Sampson was a little bit apprehensive for a young protégé of one of the great of British astronomy. Another point was that De Sitter waited rather long to write the letter. Only on 10 August 1900 he wrote with the explanation that he had wished to have more results from his calculations first. Problems were that he had to wait for results from the Cape, and the complete uselessness of Damoiseau's tables. But there were no better ones. In the land of the blind the one-eyed man is king, De Sitter wrote. He had written an extensive summary of his calculations up to then, and what his further plans were. He asked Sampson not to publish anything before his dissertation would have been published. Further De Sitter wished to know everything about the observations of Pickering, director of Harvard College Observatory, and about Sampson's plans with them. Without his consent he would of course use nothing of it. And as a last idea De Sitter proposed to combine their results in the future in making ephemerides. He wrote to Sampson on an equal footing. In reality De Sitter had still to prove himself as a scientist, he had not yet defended his dissertation, while Sampson was already a settled professor. This attitude was typical of De Sitter, he was not modest, but self-confident. He could have acted more carefully. Sampson answered on 4 November 1900. He took it very seriously and wrote an answer of eleven pages. He analysed De Sitter's letter comprehensively and gave his own data. He concluded that there was only on a few points agreement between his results and those of De Sitter. He assigned little value to such a minimal confirmation. He was adamant is his further judgement. The probable margins of error given by De Sitter were absolutely no indication at all for the accuracy to be assigned to the calculated results. And to rub more salt into the wound, considering the substantial difference between their results, those of De Sitter could in no case be accepted instead of his own. He finished somewhat more friendly and asked De Sitter to correct him, if he had misunderstood things. Sampson's letter must have shocked De Sitter. Perhaps Sampson was a bit short-tempered by the bold letter with propositions by De Sitter, but he could have written his opinions with a bit of understatement. De Sitter answered like a gentleman. His results were, with the use of further data from the Cape and an appreciable number of extra calculations, after all in the region of those of Sampson, at any rate within the margins of error. Before De Sitter's doctoral degree they exchanged a few more letters and it seemed that Sampson's criticism had abated [DUL, to/from Sampson, 10-8, 14-11-1900, 11-2, 2-4, 25-4, 3-5-1901]. In the end the row in *The Observatory* was settled. Sampson wrote to Gill in 1903 that De Sitter had taken the initiative and had proposed to cooperate in order to finish the work on Jupiter's moons. He was glad that the fireworks De Sitter and he had lighted had not caused serious burns with anyone. Gill was delighted and called in his answer Sampson, De Sitter and Cookson (at the Cape also working on Jupiter's moons) a *band of brothers*, whom he wished all success [RGO, 15/131, 15-10, 17-11-1903]. The peace had been signed. In February 1905 Sampson wrote another letter to *The Observatory* about an article by De Sitter about a *striking* error in Damoiseau's tables. In the years 1905–1907 there was a bit of squabbling about the observation of the effect of libration (see the next paragraph), and in 1911 a few *notes* and *notes on notes* appeared in the *Monthly Notices of the Royal Astronomical Society* (*MNRAS*) about the time rays of light need to reach the earth from Jupiter. But it remained on a civilized and scientific level. After De Sitter's appointment as a professor in 1908 Sampson and De Sitter worked after all separately on Jupiter's moons. Sampson published in 1910 the *Tables of the Four Great Satellites of Jupiter*. For his tables he had only used the Harvard observations. He had every right to do that, in De Sitter's opinion, who was in the meantime busy with his new analytical theory.

In his History of the Royal Observatory, Cape of Good Hope (Gill 1913) Gill pointed out that the measurements of Sampson and De Sitter differed appreciably, even more than the probable errors. A few of the data (the inclinations of the orbits of the four satellites and their nodes) from Gill's book are given below.

Inclinations and Nodes of the four Jovian satellites

I	De Sitter	0.0272 ± 0.0028	and	60.2 ± 7.0
	Sampson	0.0327	,,	33.3
II	De Sitter	0.4683 ± 0.0016	,,	293.16 ± 0.19
	Sampson	0.4644	,,	290.55
III	De Sitter	0.1839 ± 0.0026	,,	319.71 ± 0.52
	Sampson	0.1970	,,	320.70
IV	De Sitter	0.2536 ± 0.0023	,,	11.96 ± 0.67
	Sampson	0.2635	,,	7.33

Epoch 1900, Jan. $0^d.0$ G.M.T. (in degrees $°$)

Later Sampson and De Sitter corresponded now and again, but it never became heartily. Even Gill could not pour some oil in the machine. De Sitter complained about incomplete information [RGS, DOG/159, 25-4, 9-7, 8-8, 17-8, 21-9, 24-12-1910, 5-1-1912]. They were of course in a sense dependent on each other. On the one hand it is a pity that the two big Jupiter men did not cooperate. Probably their characters did not allow it. Although, in view of the differences in the above table it would have been profitable to co-operate. On the other hand, Sampson used only existing theory and mainly the Harvard observations to arrive at his tables, while De Sitter was able to develop a new analytical theory. De Sitter and Sampson met several times at congresses and meetings during their further careers, also at the meetings of the International Astronomical Union. During the first general assembly in Rome in 1922, long after the mutual commotion, De Sitter and Sampson and their wives drank a pleasant cup of tea together. Perhaps the two men will have laughed about their fireworks from the past (De Sitter-Suermondt 1922).

Libration

In his dissertation De Sitter already treated the effect of *libration* of the three inner moons of Jupiter: Io, Europa and Ganymedes. The effect arises by the special ratio of their orbital periods: $T_1:T_2:T_3 = 1:2:4$, numbered from the inner moon outward. We have then so-called commensurability, ratios with small whole numbers. The three moons are exactly in line with Jupiter once every period T_3. Laplace had deduced for these conjunctions that the angles of the orbital radii, measured from a fixed direction (called the lengths l_1, l_2 and l_3), satisfy the formula: $l_1 - 3l_2 + 2 l_3 - 180° = 0$, in which θ is small and changes periodically around the value 0. The 180° in the formula is related to the fact that when the satellites 2 and 3 are aligned on one side of Jupiter (take $l_2 = l_3 = 0$), satellite 1 is exactly on the opposite side of the planet ($l_1 = 180°$). The small variation is called libration (from the Latin *libra*, scales).

Besides this the following effect occurs. Without the ratio 1:2:4 the mutual disturbing forces of the moons on each other are to such a degree irregular in time, that they extinguish each other more or less. They leave only small perturbations, which can be described accurately in the mathematical description using a series expansion with only a few terms. The rest of the terms may be neglected. But with this particular ratio 1:2:4 the effects of the perturbing forces enhance each other at each returning identical position. Then resonance occurs, so that much more terms in the series expansion have to be calculated.

Delambre had not been able to prove the existence of the libration from the in his time available observations. Successive astronomers had therefore not taken the possible effect in their calculations. However, De Sitter had calculated in his dissertation the lengths of the three moons separately. In Chap. 3 he deduced from the lengths: $\theta = +0°, 05270 \pm 0°, 0138$. The value was four times the estimated error. De Sitter concluded that the libration effect probably did exist and could be measured. This discovery was a success, which neither Gill nor De Sitter had expected, although they had perhaps hoped to find it. But the proof was not yet very convincing. There was only a single value of θ, at a single time (the short observational period at the end of 1891, assumed that the libration period would be much longer. So De Sitter could not yet determine the period. To that end measurements were needed over a period at least as large as that period. If he wished to prove his partial discovery with certainty, further research was needed, after which the libration period could be ascertained as well. Otherwise stated, new measurements were needed, or old useful measurements. De Sitter used both kinds. He took old measurements: from the observatories in Helsingfors and Pulkowa from the period 1891 to 1898. And very recent and new measurements: heliometer and photographic measurements from the Cape Observatory from 1901 to 1904. De Sitter used the photographic plates received from the Cape in those years, not only for his major article from 1906 on the inclinations and nodes of Jupiter's moons (see Chap. 6), but also to prove beyond doubt the existence of the libration and to determine its period. But regularly he was overcome by doubts. He asked for more photographs and a little later he was more positive. He wrote a few times to Gill in 1906:

I am not at all certain about the libration.

The number of plates for a number of oppositions is very small.

I get fine results.

[RGO, 15/130, 17-4, 8-5, 6-6-1906]

But also in the theoretical field (theory and practical measurements were always connected in the calculations) he had sometimes to go through rough patches to get a thorough understanding of libration:

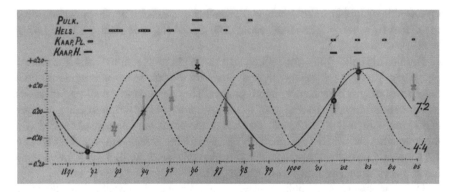

Fig. 5.4 The libration diagram, as drawn by De Sitter. Original in archive of the author

> Groping in the dark to find my way in a hopelessly complicated mass of contradictory
> results.

He had to redo or adapt much of Souillart's work. He called the new theory he
was developing the *differential theory* [RGO, 15/132, xx-3-1907]. In 1907 De Sitter
published a comprehensive article about the libration at last, with his results (De
Sitter 1907). He corrected a number of smaller errors in the calculations in his
dissertation and reached a new value for θ: θ = 0°, 1568 ± 0°, 0125. In a footnote
De Sitter wrote that *The Observatory* of May 1906 had published that his Jupiter
rival Sampson had measured the libration from photographic eclipse measurements.
He remarked that up to that time there had not followed a publication by Sampson.
De Sitter himself had already discussed the probable existence of libration in a
lecture for the Dutch Physical and Medical Congress in 1905, in which he had
mentioned the necessity to incorporate the period of libration as an unknown in the
calculations (De Sitter 1905). By these remarks De Sitter probably wished to
ascertain his priority of the discovery of the libration. In De Sitter's article of 1906 a
list appeared of ten series of measurements he had analysed, leading to 10 values of
θ over a period of 13 years, from the first in 1891 until the last in 1904. He could
deduce from these data a period of 7 years and an estimated maximum value of the
roughly sinusoidal development of 0°,16.

De Sitter's Libration diagram
In the diagram De Sitter drew his results of the libration angle θ from ten
'points' in time, determined from 10 sets of measurements over an extended
period of fifteen years. He managed to draw a sinusoidal curve through them
with a period of 7.2 years. The dotted line giving a period of 4.4 years seems
to have been gambled from only a selection of the points.

Libration data of De Sitter
All data are in degrees..

Epoch	Δl_1	Δl_2	Δl_3	θ_0
1891.75	+0.100	+0.065	−0.031	−0.157
1892.82	+0.065	+0.031	−0.020	−0.068
1893.98	+0.130	+0.018	−0.041	−0.006
1895.08	+0.117	+0.006	−0.025	+0.049
1896.15	+0.143	−0.014	−0.001	+0.181
1897.27	+0.105	+0.024	−0.014	+0.006
1898.30	+0.176	+0.106	−0.001	−0.143
1901.61	+0.136	+0.020	−0.038	+0.001
1902.60	+0.134	−0.025	−0.023	+0.164
1904.89	+0.234	+0.048	−0.004	+0.084

The data are taken from the table in De Sitter's article on libration from 1907. The epochs are decimal years. 1891.75 means 1 October 1891, the short period of the measurements of Gill and Finlay. The values in the columns 2, 3 and 4 satisfy Laplace's formula:
$\theta_0 = \Delta l_1 - 3\Delta l_2 + 2\Delta l_3$. In the column of θ_0 we see that between the two minimum values are roughly seven years. The same holds true for the time difference between the two maximum values (De Sitter 1907).

The conclusions that had to be drawn according to De Sitter from his work on the libration were twofold. First he stated at the end of his 118 page article that Souillart's theory, although an admirable work, did no longer meet the necessary requirements. In particular it was the commensurability of the orbital times that played tricks on Souillart's theory. The expansion series in the regular perturbation theory converged much too slowly by the resonance effect. In the second place De Sitter stated that a new theory was highly necessary. A new theory had to have an *intermediary orbit* as a starting point. This orbit had to have incorporated already the effects of the commensurability and of the oblateness of Jupiter. Moreover, the plane of Jupiter's equator should have to be taken as the fundamental plane, and not the plane of Jupiter's orbit around the sun. Starting from this periodic orbit the other perturbations could be dealt with and calculated. The convergence, reaching the nearly exact values, could go then much faster, with less terms. The article was the elaboration of the first indications for the existence of the libration in his dissertation. It was a success, although new observations in later years created new problems. For instance, De Sitter wrote in 1916 to Hough, then director of the Cape Observatory, that analysis of recently made photographic plates had not yielded libration [RGO, 15/132, 8-1, 18-2-1916]. The libration would appear to be much smaller on the basis of later observations, also by other astronomers, too small to determine according to some. However, his study of the libration and the comparison of it with the prevailing theory of Souillart had strengthened De Sitter in his will to create the new theory, that would improve not only the libration results but in particular those of all orbital elements of the moons of Jupiter. In 1909 he would as a first step on his way to a new mathematical theory publish the periodic solutions of the four-body problem (Jupiter and its three inner satellites. See later chapters).

Gill's Last Years

Gill, who had retired in 1907, had moved to London. There he had set himself the task to publish all his own results and those of his pupil De Sitter and assistant Cookson completely in the *Annals of the Cape Observatory*. The complete set of articles appeared in the *Cape Annals* in 1915. Gill was also busy writing his *A history of the Royal Observatory Cape of Good Hope*, which would appear in 1913. He asked De Sitter, who had already been appointed to a professorship in Leiden, to draw up plans for new series of observations to complete the knowledge of the constants in the theory of Jupiter's satellites. De Sitter answered with a firm report with proposals. Gill asked De Sitter to process all corrections on De Sitter's original dissertation, already 12 years ago, in a new publication, also with the most recent data. Again all those corrections on his dissertation, De Sitter must have cursed it. Gill wanted to incorporate De Sitter's ideas and to ensure that the observations would be made. In 1913 De Sitter again received photographic plates from the Cape [RGS, DOG/159, 27-3, 3-4, 11-4-1912, 4-8-1913]. De Sitter took care of the publication, but Gill himself wished to write the introduction. In that introduction he wrote that the original idea for the study of Jupiter's satellites arose already in 1879, shortly before his departure to the Cape Observatory, during a discussion with John C. Adams (1819–1892). Adams had predicted the existence and position of the planet Neptune on the basis of unexplicable disturbances in the orbit of Uranus. However, he had to leave the honour for the prediction to the French astronomer Urbain Le Verrier (1811–1877), who had made the same calculations, but who had found the German Johann G. Galle (1812–1910) willing to make the crucial observations. With the help of his student Heinrich L. d'Arrest Galle found the new planet in 1846 within a few days after having received the co-ordinates from Le Verrier. Adams wished to do measurements on the satellites of Jupiter in order to extend Damoiseau's tables to years after 1890. He intended to do it by measurements of the shadows of the moons on the planet when passing in front of it. Gill's opinion was that he could better perform heliometer measurements of the mutual angles of the satellites. It took twelve years till Gill and Finlay made these measurements.

On the evening of 15 December 1913, while writing his introduction to the series of articles for the *Cape Annals* Gill fell ill. He did not recover and died on 24 January 1914. His grave is in Aberdeen (Forbes 1916).

Hough, Gill's first assistant since 1898, became Gill's successor as director of the Cape Observatory in 1907. After consultation of Hough and Kapteyn de Sitter left the first part of the introduction unaltered and wrote the closing part. There was some delay by illness of De Sitter. He had to remain in bed for the whole summer of 1914 and had to have surgery in September. Hough, an excellent mathematician, did not have Gill's charisma, but under his guidance, still in the shadow of Gill, a comprehensive programme of observations was continued, during which he regularly did observations for De Sitter (Warner 1979).

In the year World War I broke out, in The Netherlands that was the only subject in discussions, De Sitter wrote. Half a million of Belgians, fled via Antwerp to The Netherlands, had to be clothed, fed and accommodated. The war, in which The Netherlands stayed neutral, had only just started, but De Sitter already thought about the situation after peace would have been concluded. The scientists should set a good example and restore the international contacts, which were necessary for the advancement of science and the well-being of humanity [RGO, 15/132, 31-1, 18-3, 12-10-1914, 19-2, 11-12-1915 to and from Hough].

References

De Sitter, W. (1901). 'The orbits of Jupiter's satellites'; *The Observatory*, Vol. 24, pp. 341–345, 1901.

De Sitter, W. (1902). 'Jupiter's mass'; *The Observatory*, 25, p. 166, 1902.

De Sitter, W. (1905). 'On the Libration of the Three Inner Large Satellites of Jupiter, and a new method to determine the mass of Satellite 1' (Dutch: 'Over de Libratie der drie binnenste groote Satellieten van Jupiter, en eene nieuwe methode ter bepaling van de massa van Satelliet 1'); *Transactions of the Tenth Dutch Physical and Medical Congress* (Dutch: *Handelingen van het Tiende Nederlandsch Natuur- en Geneeskundig Congres*), pp. 125–128, 1905.

De Sitter, W. (1907). 'On the Libration of the Three Inner Large Satellites of Jupiter'; *Publication of the Astronomical Laboratory at Groningen*, 17, pp. 1–119, 1907.

De Sitter-Suermondt, E. (1922). *Travel report Rome, on the occasion of the first General Assembly of the International Astronomical Union*, 1922. In the family archive of Tjada van den Eelaart-de Sitter, Arnhem.

Forbes, G. (1916). *David Gill, Man and Astronomer*. See References of chapter 4.

Gill, D. (1913). *A History of The Royal Observatory, Cape of Good Hope*, Published by His Majesty's Stationery Office, London, Printed by Neill & Co, Edinburgh, 1913.

King, H.C. (2003). *The History of the Telescope*, Dover Publications Inc., Mineloa, New York, 2003 (First published: Charles Griffin & Co. Ltd., High Wycombe, Buckinghamshire, England, 1955).

Mudry, A. (2005). *Galileo Galilei, Schrifte, Briefe, Dokumente*, ALBUS im VMA-Verlag, Wiesbaden 2005.

Sampson, R.A. (1901a). 'The orbits of Jupiter's satellites'; *The Observatory*, Vol. 24, pp. 271–274, pp. 301–305, 1901.

Sampson, R.A. (1901b). 'Mr. De Sitter's determination of Jupiter's mass'; *The Observatory*, Vol. 24, pp. 376–379, 1901.

Warner, B. (1979). *Astronomers at the Royal Observatory Cape of Good Hope*. See References of chapter 4.

Chapter 6
The Road to Professor in Leiden

At last my fate has been decided.
De Sitter in a letter to Gill [RGS, DOG 159, 6-8-1908]

Jacobus Cornelius Kapteyn (1851–1922)

Kapteyn was born in the village of Barneveld in the province of Gelderland, where his parents had a boarding school, where French was the official language. Kapteyn's father came from a true teachers' family. The busy parents had not much attention for their own children. Kapteyn, ninth child of fifteen, suffered from this lack of attention. He made his first astronomical observations with a telescope, a present from his father. He was very talented and went to Utrecht in order to study mathematics and physics in 1868, where he took his doctoral degree *magna cum laude* with his advisor Cornelius H. C. Grinwis (1831–1899) in 1875. The title of his dissertation was *Research into oscillating flat membranes.*[1] Kapteyn married Catharina Elisabeth Kalshoven. They had two daughters and a son. Daughter Henriëtte married the famous Danish astronomer Ejnar Hertzsprung, who would become deputy director of Leiden Observatory in 1919, when De Sitter had become director. After his doctoral degree Kapteyn became an observer at Leiden Observatory, when Hendrik van de Sande Bakhuyzen was director. Kapteyn accepted the newly established professorship of astronomy in Groningen with the inaugural lecture *The parallax of the fixed stars*. This was a first step on the way to his life's goal: to gain an insight into the structure of the universe. He was a theoretician, but also a talented observer. Because he did not receive money to equip his own observatory, he made parallax observations in Leiden during the summer holidays with the help of the meridian circle. But in this manner only few parallaxes and distances of stars could be measured with great precision. In order to obtain information of many more stars Kapteyn switched to a statistical method. After a few difficult years without instruments, in a letter of 16 December 1885

---

[1]Dutch: *Onderzoek der trillende platte vliezen.*

© Springer Nature Switzerland AG 2018
J. Guichelaar, *Willem de Sitter*, Springer Biographies,
https://doi.org/10.1007/978-3-319-98337-0_6

Kapteyn offered to Gill his support to measure photographic plates of the stars in the southern sky, although both brothers Van de Sande Bakhuyzen advised against it. Kapteyn's generous offer and Gill's immediate acceptance resulted from the fact that both men had already got to know and learned to appreciate each other by correspondence. They met for the first time in 1887 (Forbes 1916). Kapteyn did the measurements for the *Cape Photographic Durchmusterung* or CPD with the help of a by himself designed instrument. He was supported over the years by his assistant T. W. de Vries. The measurements by Kapteyn continued until 1892. *Finished!*, he wrote to Gill in June. Gill's answer was:

> This is only a jubilant shout—a hurray—a God bless you my boy,—and long may you go on and prosper. (Hertzsprung-Kapteyn 1928)

The results, the data of more than 450,000 stars, appeared in three robust volumes of the Cape Annals, the last one in 1900 (Gill and Kapteyn 1896–1900). Kapteyn focused his attention on the determination of the proper motions of stars. In 1896 he realized that the properties of the 2400 stars he had measured were not, as assumed, randomly distributed over all directions. In 1904 he established that there were two preferential directions, or *star streams*, later called by Eddington one of the five greatest astronomical discoveries of the nineteenth century. The chief assistant of Hough at the Cape Observatory, the German Jacob K. E. Halm (1866–1944), discovered a third stream in 1915 (Warner 1979). Kapteyn made known his astonishing discovery in his lecture at the St. Louis World Fair and International Congress of Arts and Sciences Exposition in 1904. He was invited by the Canadian–American astronomer Simon Newcomb (1835–1909).

Fig. 6.1 Jacobus C. Kapteyn. Photograph from (Hertzsprung-Kapteyn 1928)

In the ensuing years Kapteyn came with his *Plan of Selected Areas*, the next step in the solution of the structure of the universe (DeVorkin 2000). It was necessary to measure a large number of data of as many stars as possible: magnitude, proper motion, parallax, spectral class and radial velocity. Because measurements on the complete sky would be too much work, Kapteyn chose for the stars in 206 selected small areas in the sky. It became a real international enterprise, in which a lot of observatories cooperated for years. The radial velocity of a star, in the direction from the earth or to the earth, which could be measured with the help of the Doppler effect accurately, gave, together with the proper motion (perpendicular to the radial velocity), an indication of the average distance of stars. Kapteyn had to make use of statistical methods in his analyses. It was not about individual stars, but about position and velocity of groups of stars, for which he was able to calculate average values.

Doppler effect

In the electromagnetic spectrum of a star, the radiation we receive, we may discern absorption lines, sometimes groups of lines, belonging to a special element, that we know from experiments on earth. The wavelength λ' of such a spectral line that we receive on earth, can be larger than the same wavelength λ we measure in a laboratory on earth. This is a result of the fact that the star is moving with a velocity v away from the earth. Suppose that a complete wavelength is sent from the star. The front of the wavelength travels to the earth with the speed of light c. When the last tiny piece of the wavelength is leaving the star, it is a small distance further from the earth. Thus the wavelength λ' speeding to the earth with the speed of light, is larger than λ, measured in a laboratory on earth. From the difference between λ' and λ we can calculate v. But we must keep in mind that it only is the radial component v_r of the star's velocity relative to the earth.

On the basis of the observations during the years following the launch of his plan, Kapteyn wrote in 1922 his First Attempt at a Theory of the Arrangement and Motion of the Sidereal System (Kapteyn 1922). Two years earlier he had published provisional results with Van Rhijn.[2] Before his article in 1922 he had already talked about it at the meeting of the British Association in September 1921 and in November at Leiden Observatory. It was a magnificent conclusion of his life's work, after the first tentative attempt by William Herschel to get some insight in the form of our Galaxy at the beginning of the nineteenth century. The name Kapteyn Universe has been connected with it since. In the end our own Milky Way appeared much larger than Kapteyn thought (interstellar dust limited the sight) and all nebulae appeared to be galaxies on their own. However, his epoch-making work must

---

[2]Pieter J. van Rhijn (1886–1960) was a student of Kapteyn and became director of the Astronomical Laboratory Kapteyn in Groningen in 1921.

not be underestimated, in view of the fact that forty years earlier, when he started his investigations, hardly thirty parallaxes of near stars were known. His student Jan Oort,[3] whom De Sitter persuaded to come to Leiden in 1924, followed in his footsteps and made use of statistical methods in his later calculations on the structure of our Milky Way. Oort would become director of Leiden Observatory in 1945.

In 1921 Kapteyn retired and his Astronomical Laboratory was renamed in Astronomical Laboratory Kapteyn, later Kapteyn Astronomical Institute. Van Rhijn succeeded him in Groningen and continued working on the Selected Areas.

Ko and Elise, the forenames of Kapteyn and his wife, had a harmonious and happy family life and were loved by all around them. They were hospitable and loved children, as Bertha Clemens Schröner, a girl next door in Groningen around 1920, still remembered very well in 2008, at the age of 97.[4] Kapteyn and Gill and their wives were good friends. When Gill had returned home in 1896 after the visit to Groningen, where he had trapped De Sitter into coming to South Africa, he wrote to the Kapteyns about their daughters:

> I had a lovely time in Groningen. I will write my little sweethearts in a couple of days.
> (Forbes 1916)

Kapteyn's daughters found that silly and sociable English uncle rather funny (Hertzsprung-Kapteyn 1928). George Ellery Hale (1868–1938), director of the Observatory on Mount Wilson in Pasadena near Los Angeles, also became a good friend. From 1908 until 1914 Kapteyn, accompanied by his wife, stayed there as a *Research Associate*, each year for a few months. At first in a tent (they did not care for status), later in the Kapteyn Cottage. Kapteyn died in 1922. De Sitter's friend and American astronomer Frank Schlesinger (1871–1943) wrote to him:

> The world has lost its greatest astronomer.
> [ASL, 28-7-1922]

De Sitter and his friend Huizinga, professor of history, took on themselves the task to write a biography of Kapteyn. Hale wrote to do his best to collect material. He had once asked Kapteyn to write down all his future plans, but he was afraid Kapteyn had not come very far [ASL, 28-1-1925]. A request from De Sitter appeared in *The Observatory* in September 1925 to send him documents about and reminiscences to Kapteyn. A lot of relevant documents and correspondence was sent from Groningen to Leiden in a sizeable chest. The Dutch astronomer Adriaan Blaauw, in the early 1930s a young student in Leiden, was in later life under the impression to have seen this chest in the attic of the Observatory. The writing of the Kapteyn biography did not happen for a number of reasons. One of them is that the chest got lost. One theory is that the chest ended up in the harbour of Rotterdam.

---

[3]See for further information about Oort the biography of Van Evert (Van Evert 2012).
[4]Private communication from Mrs. Bertha Clemens Schröner (1911–2009). She remembered calling them Kappie and Ootje.

Perhaps, after De Sitter's death, the idea had come up that De Sitter's son Aernout, appointed director of the Bosscha Observatory in Lembang in the Dutch East Indies in 1939, would take it with him. The chest would then have been left behind there, because it contained the Kapteyn papers. After that it could have been destroyed during the bombardment of Rotterdam in May 1940. There probably will never be any certainty about the fate of these Kapteyn documents (Van Maanen 1922; Sears 1922).

It took 93 years before the complete biography of Kapteyn was published, a major accomplishment by Piet van der Kruit (since 1975 professor at the Kapteyn Astronomical Institute in Groningen) and a *Fundgrube* for anyone, astronomer as well as historian, interested in Kapteyn and his work (Van der Kruit 2015).

Assistant and Private Lecturer with Kapteyn

Kapteyn was in doubt about the future direction of De Sitter's astronomical work in 1902, and he asked Gill for advice. Was it sensible for De Sitter to keep on working at the satellites of Jupiter? On the one hand he needed De Sitter for other work. But on the other hand, if a spectacular aim could be reached with it, Kapteyn did not want to prevent it. But if it would appear later that De Sitter should have wasted his time on second-class work, he would not easily forgive himself. Gill left the decision to Kapteyn [RGO, 15/131, 11-3, 18-10-1902]. After the somewhat hectic correspondence with Sampson in *The Observatory* about inaccuracies in De Sitter's work, De Sitter himself sometimes had doubts about his further career. In the meantime, after the birth of a son, Lamoraal Ulbo, the De Sitters had a family with two small children. De Sitter had a small income and life got more expensive. He was in doubt, but in the end decided, with Kapteyn's consent and to Gill's satisfaction, to continue working on Jupiter's satellites. At the Cape Observatory new series of photographs of Jupiter and its moons were made. On those plates the positions had to be measured. In fact all according to the plans De Sitter had set out in chapter IV of his dissertation. Kapteyn had done his best and an allowance from the University Fund was awarded. The money was intended for the determination of the nodes and inclinations of the moon orbits and of the compression or flattening of Jupiter. Now he had made his decision, he was eager to receive the available data. If the photographic plates had not yet been measured De Sitter himself could also measure them with the Repsold machine. De Sitter also wanted to take into his research the plates still to be taken in 1903 and 1904. But for a start he wished to restrict himself to the inclinations and nodes. The rest he left to Sampson for the time being. In the meantime De Sitter worked on measurements of the colours of stars. He found no systematic differences between the stars in the Milky Way as a whole and those at the *galactic poles* (the direction perpendicular on the galactic plane). However, he had discovered areas in the Milky Way with stars that were on average bluer, and areas at the poles with stars that were on average redder. De Sitter thought it resulted from the effects of absorption by cosmic dust clouds. Gill was very interested and wrote about photographs with long exposure times, on which between the weak stars sharp oval formed areas without stars could be seen.

Galactic co-ordinates
In order to position a star sometimes galactic co-ordinates are used. The earth is the centre, and the main direction is the line from the earth to the centre of the Galaxy. The direction of a star may then be given by:
** the galactic longitude l, and
** the galactic latitude b.
The galactic longitude l is the angle in eastern direction between the main direction and the direction from the sun to the point in the galactic plane directly under or above the star.
The galactic latitude b is the angle between the direction from the sun to the star and the galactic plane.

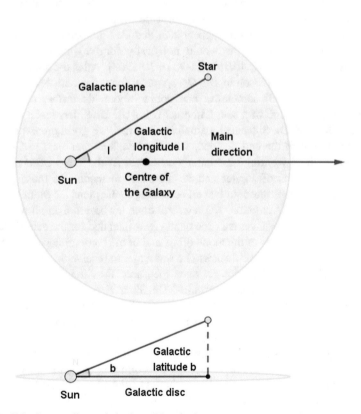

Fig. 6.2 Galactic co-ordinates latitude and longitude

On the path of theory De Sitter took his first steps in this period. He had taken up the study of Poincaré's recent books on celestial mechanics. The French mathematician and theoretical physicist J. Henri Poincaré (1854–1912) had published the three-volume *New Methods in Celestial Mechanics* (*Les Méthodes Nouvelles de la Mécanique Céleste*) in the 1890s, in which he made great progress in the theory by an extensive study of the three-body problem. In particular he was interested in the theory about the periodic solutions of the more-body problems. In the existing theory there were a number of uncertainties.

Souillart had showed that Laplace in the theory of Jupiter's satellites had neglected terms erroneously. But there were also doubts about the terms neglected by Souillart himself. De Sitter saw possibilities to make better series expansions. He made his first cautious attempts to reach a new theory for Jupiter's moons. It took more than a decade before he would publish the final results. De Sitter wrote to Gill about his ideas. He also wrote that his children, of who he included a few photographs, enjoyed life, like the parents. Willem and Non had been able to overcome the sorrow of the loss of their first child [RGO, 15/130, 29-6, 5-8-1902].

Gill visited England again in 1904, and also Groningen. De Sitter would be glad to have his advice about his research into the nodes and inclinations. Which values would he use for the other orbital elements? He sent his letter to the Athenaeum Club at Pall Mall in London, of which Sir David, as a man of great scientific distinction, was a member as a matter of course [RGO, 15/130, 28-1, 22-6-1904]. A substantial number of plates had been made at the Cape during the oppositions of Jupiter in 1903 and 1904. And in 1905 De Sitter wrote that he had nearly finished his work on them. In the meantime De Sitter had discovered yet another number of mistakes in his dissertation. He wrote:

Fig. 6.3 J. Henri Poincaré. Photo from Deutsches Museum München

I wish I had never written it, so hopelessly many mistakes.

But the only thing he could do was to correct them [RGO, 15/130, 14-2-1905].
In June 1905 he poured out his heart in a long letter to Gill.

I am certain that you made few mistakes in your life. But one you did make: that you
entrusted me with your beautiful heliometer measurements. It was then a far too difficult
task for me. I blundered horribly when working with them.

De Sitter had been a few times on the verge of giving up astronomy completely.
He wrote a list of his mistakes. A number of them are: the correct part of Sampson's
criticism; the missing out of the aberration[5] of the fixed stars; the not taking into
account the difference in distance to the earth of two moons, resulting in the fact
that two light rays leaving the moons at different times do arrive at the earth at the
same time. In his own words De Sitter had learned to be very cautious with
statements about the absence of mistakes [RGO, 15/130, 20-6-1905]. The somewhat overconfident young scientist had meanwhile eaten a bit of humble pie. But it
seems improbable that his thoughts to abandon astronomy ever became further than
an impulse, caused by a temporary annoyance. His fiery desire to achieve something splendid was unquenched. He received much support: from Kapteyn, from
Gill and from his wife. They had a happy family life and in April 1905 they had
another son: Aernout. But besides the annoyance about his omissions, there were
enjoyable events as well. The congress of the British Association for the
Advancement of Science (BA) was organized in South Africa in the fall of 1905.
Officially Gill had approached the nestor of Dutch astronomy Van de Sande
Bakhuyzen to send him five names of excellent Dutch scientists to be invited, for
who all costs would be paid [RGO, 15/130, 1-4-1904]. Kapteyn and De Sitter
received an invitation [RGO, 15/130, 25-1, 7-6, 20-6-1905] and they set sail to
Cape Town on the Kildonan Castle. The group of astronomers aboard established
the Astronomical Society of the Atlantic for the duration of the voyage (Van der
Kruit 2015). For De Sitter it was a welcome reunion with his beloved South Africa
and with the friends he had learned to know there six years ago. After the not
spotless period of writing and reception of his dissertation in the scientific world, it
was an honour to be invited to a big congress. The British Association was founded
in 1831, a bit as a protest against the elitist Royal Society, and always had made
great efforts to draw attention to the sciences with the general public. At the
congress Kapteyn lectured, after his announcements in St. Louis in 1904, again on
star streams (Kapteyn 1906).

The return trip passed rather unlucky. The steamer Durham Castle was moored
up for days in the Suez Canal. Many passengers left the ship in order to travel home
via Italy. De Sitter used the free time to visit the pyramids. During the visit of a rock
grave he fell into a four meter deep hole and was severely wounded above his left
eye. He had to stay in hospital for two days for treatment.

---

[5]Aberration of light. As a result of the moving of the earth the light of a star seems to come from a
slightly different direction, compared to the direction with a not-moving earth.

In the beginning of 1906 De Sitter started to write his article on the nodes and inclinations, which would appear in the Annals of the Cape Observatory. He sent it, 150 pages long, to Gill in April, who could let it appear in print unaltered. For the record De Sitter wrote that this article, as promised by Gill, would appear under his own name (De Sitter 1906). The reason was that articles often appeared under the name of the director of the observatory. But De Sitter was working on his publication list.

De Sitter thought, after this article about the nodes and inclinations and the one about the libration, to stop temporarily with the Jupiter moons. The Astronomical Laboratory required his time. The first photographic plates of Kapteyn's Selected Areas would arrive soon. De Sitter had to work on them and informed Gill [RGO, 15/130, 6-7, 1-8, 24-8-1906]. In 1908, even before De Sitter's appointment as a professor, Gill tried to appoint De Sitter as an *Associate of the Royal Astronomical Society* (RAS). The attempt failed, but De Sitter came to know of it through Kapteyn and he wrote that he would perhaps be worth it in the future, but that it would now be out of proportion, in view of his humble qualities. In 1909 Gill, then president of the Royal Astronomical Society, did the same proposal, now with success.[6] De Sitter was proud [RGS, DOG/159, 24-4-1908, 23-5-1909]. This membership would give him a direct contact with a lot of English astronomers. He became a regular visitor of the meetings of the RAS in London. These English connections of De Sitter were a direct consequence of those of Kapteyn and they would always be of great use to him during the rest of his career.

Membership of the Natural Sciences Society in Groningen[7]

Besides De Sitter's own scientific work he became a member of the Natural Sciences Society. This Society existed in Groningen since 1801. During the seventeenth century there was a growing interest in the natural sciences, leading to the start of the modern natural sciences. Two of the main scientists of those early scientific period were the mathematicians, physicists and astronomers Isaac Newton (1643–1727) from England and Christiaan Huygens (1629–1695) from The Netherlands. Scientific organizations were founded, like the Royal Society in England. During the eighteenth century the bourgeoisie became also interested in the new sciences. Instruments like vacuum pumps, telescopes and planetaria were widely used in demonstrations. The first scientific Society in The Netherlands was founded in 1752: The Holland Society of Sciences (see next chapter). More of them followed. In Groningen the Natural Sciences Society was founded in 1801 by Theodorus van Swinderen (1784–1851) and a small group of co-founders. Since then more than 1800 lectures have been organized by the Society, which is alive

---

[6]*MNRAS*, LXIX, 8, p. 615.
[7]Dutch: Natuurkundig Genootschap. Since 1975 Koninklijk (Royal) Natuurkundig Genootschap.

and kicking up to the present day. Under the chairmanship of Florentius G. Groneman (1838–1929), director of the hbs in Groningen the Society went through a golden age. In 1899 a scientific department was founded with the aim to organize, alongside the lectures for the interested bourgeoisie, lectures for the interested listeners with a better knowledge of the sciences. After a few years Kapteyn became chairman and De Sitter secretary of the scientific department of the Society. Already in 1878 Kapteyn held his first lecture for the Society about *The relation between the galaxies*. On a regular basis De Sitter also gave lectures for the Society on subjects from his investigations. In those years Kapteyn and De Sitter met the original Frisian physicist Obe Postma (1868–1963), who took his doctor's degree with the Dutch physicist and Nobel prize winner Johannes Diderik van der Waals (1837–1923). Postma lectured a number of times for the Society. Nearly his whole career he was a teacher of mathematics at the hbs of Groneman, from 1894 until 1933.[8]

Postma also became one of the most loved Frisian poets, and investigator of the history of the Frisian country and life. After his retirement he moved to Leeuwarden and stayed active until his death at the age of 95. For more information about the Society see Smit (2001) and Guichelaar et al. (2012).

Postma must have been impressed by De Sitter, because after his death in 1934 he wrote a poem for De Sitter. In translation[9]:

On the Occasion of the Death of a Great Astronomer
Their ways reach into infinite spaces,
Their senses know what remotest worlds are like,
Their thoughts do not falter when our understanding fails,
And harmony appears [to them] where no order reveals itself [to us].
To the poet beauty comes, mild as heaven,
The seeker gains the cosmos, the ordered image.
The Divine being/essence—are they outside of him?
Tends to enter our world, and speaks to humans.

Other Publications Until 1908

Colour of stars depending on galactic latitude

While working on the CPD Kapteyn conjectured that there was a difference between the visual and photographic magnitudes of stars, depending on the fact if

---

[8]The hbs in Groningen was the first in The Netherlands where the specific curriculum was taught, in 1864. The Groningen hbs was founded with the support of Willem de Sitter, the grandfather of the astronomer and in those years mayor of Groningen. Groneman was director of the hbs from 1869 until 1905. His scientific motto was: *No truth ever stood in the way of another one.*

[9]The translation is made by the Frisian poet and translator Geart van der Meer, to whom I am very grateful. The original poem is published in: Steenmeijer-Wielenga, T. (editor), *Obe Postma, Samle fersen*; Friese Pers Boekerij, Leeuwarden, 2005.

the stars were in the galactic plane or in the vicinity of the galactic poles. So, if two stars, one in the direction of the galactic plane and another near a galactic pole, had equal visual magnitudes, there was often a difference between their photographic magnitudes. In order to do further research into this particular effect, he had asked to make new plates at the Cape Observatory between 1893 and 1895. These were measured by De Sitter in Groningen before leaving to the Cape. He could draw the conclusion that the effect conjectured by Kapteyn was not caused by meteorological or seasonal influences. But he could not decisively answer the question whether the differences were caused by systematic errors in the visual measurements or that there really was a difference between the galactic and the polar stars. He wrote two articles about these Groningen measurements at the end of 1897, already at the Cape, that were published in 1900 (De Sitter 1900a, 1900b). De Sitter had taken with him to the Cape a Zöllner photometer to make there photographic plates of a number of the same stars. The apparatus was mounted on a 6-in. telescope. Innes supported De Sitter in this research by measuring the stars a second time visually. For the determination of the brightness of a star an *artificial star* was used for comparison. With the help of the light of an oil lamp a pseudo star was imaged through a number of prisms together with the image of the observed star. By comparison the magnitude of the star could then be determined. The first photometer of this kind was made by the German astronomer Johann K. F. Zöllner (1834–1882). The problem with the oil lamp was that it flickered too much during a strong southeasterly wind, not unusual around the Cape Observatory, which gave an unstable image. In that case it was impossible to make observations. De Sitter experimented with an electric 4-V lamp instead of an oil lamp, fed by batteries in the cellar. By fastening the lamp to the instrument, on the advice of Gill, De Sitter managed to obtain a stable image, immune to a severe wind. In this way De Sitter did his first experimental investigations and obtained also a first insight into the functioning and possible technical adaptations of astronomical apparatuses. Later, when he was director of Leiden Observatory, several new telescopes and ancillary apparatuses would be built according to his designs. De Sitter published about the Zöllner photometer in the *Astronomische Nachrichten* in 1903 (De Sitter 1903a) and in the *Monthly Notices* in 1899 (De Sitter 1899). The concluding article about the observations of Innes and De Sitter appeared in 1904 (De Sitter 1904). Again the conclusion was that they dealt with a real effect, and not a systematic measurement error. The colours of the stars at the poles of the Galaxy were randomly distributed, but in the plane of the Galaxy some colours had preferential directions. There was no explanation yet. De Sitter wrote that it was of the utmost importance to find the possible connection with the general structure of the Galaxy. His conjecture was right. Later it appeared from investigations that the effect was caused by a greater proportion of hotter stars in the plane of the Galaxy (Oort 1935). So Kapteyn had conjectured correctly in 1890. De Sitter's article of 1904 was his first major publication. As a spin-off he wrote a short article in 1903, in which he pointed to six stars that had emerged in the photographic, photometric and visual investigation as probable variable stars. Some specific variable stars (stars with changing magnitudes in time) later played an important role in determining distances of distant objects in the Galaxy and beyond (De Sitter 1903b).

Magnitude of a star

The Greek astronomer Hipparchus (190–120 BC) divided the stars into six classes of magnitude: the brightest (1) up to the barely visible (6). Much later it was decided that the magnitude of one star is 1 higher than that of another star, if its luminosity is 2.5 times lower (light/energy reception on earth.). Thus, between magnitudes 6 and 1 there is a factor of $2.5^5 \approx 100$. This magnitude, as measured on earth, is called the relative magnitude m. The absolute magnitude M is the magnitude 'measured' on earth, if the star would be 'placed' at a distance d of 10 parsec (1 pc $\approx 3.0 \times 10^{16}$ m ≈ 3.3 light-year). M is a measure of the energy emitted per second. Between the magnitudes m and M, and the distance d we have the relation: m − M = 5·log(d/10). (If the distance is 10, then we have m = M. If the distance is 10 times larger (d = 100), the light received is 100 times smaller. The formula gives then m − M = 5, corresponding to $2.5^5 \approx 100$).

Parallaxes and proper motions of stars

For the extensive investigation of Kapteyn into the structure and dynamics of the Galaxy many data of large numbers of stars were necessary, particularly distances and proper motions. The distances were mainly measured with the help of parallaxes. De Sitter produced two large publications in this field, written during his time as an assistant of Kapteyn. The first one contains the results of the measurements from photographic plates of the proper motions of 3300 stars, and the second one those of the parallaxes of 3650 stars (Kapteyn 1908a; 1908b). In his preface to both articles Kapteyn wrote that the writing and the bulk of the supervision of the calculations were De Sitter's work. Even more so this applied to the discussion of the measured data.

Thus, De Sitter's investigations as Kapteyn's assistant concerning the structure of the Galaxy, were successful. Further work in this field would have yielded no doubt important discoveries. His teacher Kapteyn would have loved to make use of De Sitter in further investigations and he would certainly have given him all the credits for later discoveries. In view of all this work it is amazing that De Sitter managed to do all these investigations beside his work on the satellites of Jupiter and his first theoretical studies based on the works of Poincaré. Although he had already shown his great talents for the experimental astronomy, in the end he chose for celestial mechanics and in particular the theoretical part of it. Only measuring and measuring was in the end not what he wanted. A new theory for Jupiter's satellites, an improvement of Souillart's theory, had been haunting his mind for years. In correspondences he had made his ideas on this subject clear since 1902. He just had to follow that route.

In the meantime his chances for a professorship became actual, when Van de Sande Bakhuyzen's retirement as director of Leiden Observatory was imminent.

De Sitter's Appointment as a Professor in Leiden

Hendrik van de Sande Bakhuyzen, director of Leiden Observatory and professor of astronomy at Leiden University had to retire in 1908 *propter aetatem* (because of his age, he had turned 70). Bakhuyzen seemed to have a plan ready for his succession. In the meeting of the faculty of mathematics and physics at the end of October 1907 he tried to induce the faculty to ask the Minister to provide a supplementary budget to establish an extraordinary professorship for theoretical astronomy. The relief for the new director of the Observatory caused by this extra position would be advantageous for the appointment of a new director. The faculty committee of preparation to fill the vacancy was composed and had the following members: Van de Sande Bakhuyzen, Lorentz,[10] Kamerlingh Onnes[11] and Kluyver.[12] In the beginning of November the faculty meeting voted about the new extraordinary professorship. The votes were equally divided: 6–6. A new committee was established: Van de Sande Bakhuyzen, Lorentz, Franchimont[13] and Martin.[14] Van de Sande Bakhuyzen informed the committee that his ideal successor Kapteyn, whom he had approached, had declined his offer. The faculty tried it a again by letter [AUL, Faculty letters, 30-11-1907], but Kapteyn wrote kindly but clearly in January 1908 that he would not come to Leiden. After another couple of meetings the faculty decided in March with a vote of 7–5 to ask for a second professorship in astronomy [AUL, Faculty minutes]. The faculty put forward in the recommendation to the Board of Governors[15] that for the position of professor of experimental astronomy only one candidate qualified: the younger brother of Van

---

[10]Hendrik A. Lorentz (1853–1928) was one of the world's leading theoretical physicists. He was professor of theoretical physics at Leiden University from 1878 until 1912, when he was succeeded by Paul Ehrenfest (1880–1933). Lorentz received together with Pieter Zeeman (1865–1943) the Nobel Prize for physics in 1902 for their discovery and explanation of the influence of a magnetic field on spectral lines.

[11]Heike Kamerlingh Onnes (1853–1926) was an experimental physicist and professor at Leiden University from 1882 until 1923. He became famous for the liquefaction of helium and the discovery of superconductivity. He received the Nobel Prize for his work at low temperatures in 1913.

[12]Jan C. Kluyver (1860–1932) was professor of mathematics at Leiden University from 1892 until 1930.

[13]Antoine P. N. Franchimont (1844–1919) was professor of chemistry at Leiden University from 1874 until 1914.

[14]Karl Martin (1851–1942) was professor of geology at Leiden University from 1877 until 1922.

[15]Dutch: College van Curatoren.

de Sande Bakhuyzen. The faculty had at its disposal Kapteyn's opinion: Van de Sande Bakhuyzen's scientific work was in his field excellent and impeccable, under his guidance the name of the Observatory would be maintained. For the theoretical astronomy the faculty had four candidates. First De Sitter, and further J. Weeder (assistant at the Observatory), J. W. J. A. Stein[16] and Pannekoek. The advice of Kapteyn was probably put in positive words for the young Van de Sande Bakhuyzen to enlarge the chances for De Sitter. We don't know what the real motives of Kapteyn were to say no to Leiden (Van der Kruit 2015). The Minister of the Interior, and Prime Minister, Th. Heemskerk wrote to the Board of Governors in May that he objected to supplementary budget for a second professor. He also asked if the young Van de Sande Bakhuyzen could become professor and De Sitter observer and lecturer [AUL, Faculty letters]. The Board of Governors, frightened by the idea of the Minister, advised against the appointment of Van de Sande Bakhuyzen, because

he utterly misses the gift to pass on his knowledge to others.

This was curious, because at the start of the procedures he would become a professor. The proposal to the Minister now became: Van de Sande Bakhuyzen director of the Observatory and not professor, and besides a professor of theoretical astronomy. The older brother was willing, probably to realize a professorship for his younger brother later, to continue the supervision of the Observatory, on the condition that money would become available in 1909 for a second (extraordinary) professor [AUL, BG minutes]. In the beginning of June Lorentz wrote a curious letter to Gill [RGS, DOG/86, 7-6-1908]. In that letter he asked Gill's advice concerning young Van de Sande Bakhuyzen, in the meantime already 60 years of age. Lorentz wrote his opinion to Gill, that he considered young Van de Sande Bakhuyzen the best astronomer in The Netherlands (except for Kapteyn, but he was not available). He was completely suitable to become a director. Lorentz went even further. He thought it in the interest of the University that the professorship would not be split and would go completely to Van de Sande Bakhuyzen. Lorentz did add that he realized that the young Van de Sande Bakhuyzen was not an eloquent man and perhaps could not hold a large audience spellbound. But, he finished:

In the first place we must put the emphasis on scientific merits.

We don't know Gill's answer. But a few questions may be asked. Lorentz must have known of the qualities of De Sitter. Kapteyn and Lorentz regularly met at the monthly meetings of the Academy of Sciences in Amsterdam. He must have known that Kapteyn was seeking a professorship for De Sitter. He must also have understood that De Sitter's qualities exceeded those of the young Van de Sande Bakhuyzen. Was it to please his old friend Hendrik van de Sande Bakhuyzen? We do not know.

---

[16]Johan W. J. A. Stein (1871–1951) was a member of the Society of Jesus. He became director of the Vatican Observatory in 1930.

Fig. 6.4 Hendrik A. Lorentz. Photograph from the Liber Amicorum for Hendrik van de Sande Bakhuyzen

A week later Lorentz, Kamerlingh Onnes and Kluyver asked to be received by the Board of Governors to make some further communications in the case of the vacancy in astronomy. On 18 June the Minister wrote to the Board of Governors to have objections to their proposal. He was prepared to appoint De Sitter as a professor and to provide Van de Sande Bakhuyzen with a financial allowance as first observer, or to appoint him as an extraordinary professor (3000 guilders yearly). But the Minister did not wish to separate the directorate from the professorship. In his opinion De Sitter would also become director of the Observatory. The Board of Governors listened to Lorentz and Kluyver and decided to make another attempt and to ask the Minister to appoint De Sitter as an ordinary professor and Van de Sande Bakhuyzen as an extraordinary professor (4000 guilders yearly), at the same time in charge of the management of the Observatory. The lobby for the young Van de Sande Bakhuyzen was strong.

In the beginning of August 1908 Gill, retired and living in London, under whose guidance De Sitter had become a professional astronomer at the Observatory of Cape Town in South Africa in the years 1897–1899, received a handwritten confidential letter from De Sitter. He wrote:

> At last my fate has been decided. I am appointed as professor and director of Leiden Observatory.

> [RGS, DOG 159, 6-8-1908]

We follow the letter further. The Board of Governors of Leiden University wished to appoint the younger brother of the retiring director, Ernst Frederik, then already 60 years of age, as director of the Observatory. And next to him they wished to create a new professorship of theoretical astronomy. But the Minister in The Hague, De Sitter continues,

> would not hear of it. At least that is the course of events as I suppose it has been, officially I know nothing.

De Sitter asked his old master advice on measuring programmes for the diverse instruments [RGS, DOG 159, 6-8-1908].

But in his excitement De Sitter had written too soon. The procedure had not yet been completed. The position of the elder brother Van de Sande Bakhuyzen was split in two after all. De Sitter became professor of theoretical astronomy and the younger brother Ernst van de Sande Bakhuyzen became director of the Observatory.

The appointment of De Sitter was made by royal decree of 27 July. In the beginning of August De Sitter received the message. He was convinced that his appointment would include the directorate of the Observatory, probably because he had heard that the Minister in his appointment letter of 1 August still had objections to appointing Van de Sande Bakhuyzen as director of the Observatory [AUL, BG minutes]. On 3 August four professors, among them Lorentz, urged the Board of Governors again that it was desirable to appoint Van de Sande Bakhuyzen to director of the Observatory. There was confusion. The case was not yet definitive. There was still discussion, also at the faculty meeting. The faculty wrote on 11 August to the Board of Governors that De Sitter wished the directorate of the Observatory to be commissioned to another person. It is not clear if pressure was put on De Sitter, he knew his strong position in view of the opinion of the Minister. It may also very well have been possible that he was glad to be able to avoid the directorate and only be appointed for theoretical astronomy. On 12 August De Sitter clarified his point of view in the meeting of the Board of Governors: he had chosen for theoretical astronomy only. And they appealed again to the Minister to agree to the separation. On 17 August the Board of Governors wrote that the Minister still wished De Sitter as director and Van de Sande Bakhuyzen as observer. The Board of Governors turned again to the Minister, together with a letter from De Sitter [AUL, Faculty papers, *Vacancy astronomy 1908*]. But in the end the Minister gave in and wrote on 7 September to agree to the separation, on the condition that clear written arrangements would be made concerning the use of the Observatory by De Sitter [AUL, BG minutes] [RGS, DOG/159, 30-11-1908 to Gill]. In a later letter to Gill, of 30 November 1908, De Sitter described his doubts in this period and his final decision not to opt for the directorship. He wrote that he himself went to The Hague and after a talk of more than an hour with the Minister he succeeded in convincing him to appoint Van de Sande Bakhuyzen as director of the Observatory and extraordinary professor (Van der Kruit 2015).

So in the end the split in two of the astronomy in Leiden was a logical solution, serving several participants.

The procedure still requires some analysis. Did the older Van de Sande Bakhuyzen, with a bit of nepotism, clear the road for his brother to become director and did he seek to that end the support of his friend Lorentz? It is understandable that Lorentz tried to acquire Gill's support for the appointment of the young Van de Sande Bakhuyzen as director of the Observatory. After all, his career as a highly qualified and experienced observer had been immaculate up to that time. But it is strange that Lorentz in his letter to Gill proposed to put the complete Leiden astronomy into the hands of Van de Sande Bakhuyzen. The fact that De Sitter towered high above Van de Sande Bakhuyzen as a theoretician, was evident to all in 1908. Perhaps some saw in him only a theoretician, unfit to lead an observatory. In this matter he had once been quite frank to Gill about his future. He had written him in 1907 that he preferred an independent position leaving him time for his theoretical work. The directorate of an observatory would be a heavy burden, he was ashamed to add. Although he would accept it, because he had to bring home the bacon. The reason for De Sitter's frankness was his deeply felt wish to develop a new theory of Jupiter's satellites, and that would require all his energy [RGO, 15/132, 15-3-1907]. Perhaps there were in the faculty ideas of expanding the number of professors in other sciences, so that two astronomers would diminish the chances for such a development.

The successful lobby for Van de Sande Bakhuyzen became not a success for himself. The Observatory reached a low in research and publications. For De Sitter all went well. The years between 1908 and 1918 were scientifically the most fruitful of his career. At the end of that period he published articles on cosmology (1916–1917), containing among other results his model of the universe, and his major articles concerning his new theory of Jupiter's satellites (1918–1919).

In the beginning of August 1908, directly after De Sitter had received his appointment letter, his parents, brother and sister came to congratulate him in Bergen at sea (in the province of North-Holland), where the family, now with the youngest daughter Agnes, born in 1907, were taking their holidays. De Sitter's father, according to De Sitter prouder of his appointment than De Sitter himself, would die in September. He missed De Sitter's inaugural lecture.

Fig. 6.5 The family came to congratulate De Sitter to his appointment as a professor in 1908. From left to right: Eleonora, Ulbo, Willem, grandfather De Sitter, Aernout, Dora, grandmother De Sitter, brother Ernst. Sister Bine was absent or took the photograph. Photograph from the family photo album of the astronomer's brother Ernst Karel Johan

The young family moved from Groningen to Leiden in October, Noordeindeplein 2B, and later to Witte Singel 37, just outside the old city moat.

After the unexpected death of Van de Sande Bakhuyzen in 1918, De Sitter became director of the Observatory after all, again after a procedure full of uncertainties.

References

De Sitter, W. (1899). 'On the Use of the Electric Light for the Artificial Star of a Zöllner Photometer'; *MNRAS*, 59, p. 341, 1899.
De Sitter, W. (1900a). 'On the systematic difference, depending on galactic latitude, between the photographic and visual magnitudes of the stars'; *Publications of the Astronomical Laboratory Groningen*, 2, pp. 5–22, 1900.
De Sitter, W. (1900b). 'On Isochromatic Plates'; *Publications of the Astronomical Laboratory Groningen*, 3, pp. 23–26, 1900.
De Sitter, W. (1903a). 'Über die Intensitätskurve bei Beobachtungen mit dem Zöllnerschen Photometer'; *AN*, 163, p. 65, 1903.
De Sitter, W. (1903b). 'Suspected Variable Stars'; *AN*, 162, p. 205, 1903.
De Sitter, W. (1904). 'Investigations of the systematic difference between the photographic and visual magnitudes of the stars depending on the galactic latitude, based on photometric observations by W. de Sitter, visual estimates by R.T.A. Innes, and photographs taken at the

Cape Observatory, together with catalogues of the photometric and photographic magnitudes of 791 stars'; *Publications of the Astronomical Laboratory Groningen*, 12, pp. 1–167, 1904.

De Sitter, W. (1906). 'A determination of the inclinations and nodes of the orbits of Jupiter's satellites by Dr. W. de Sitter. From photographic plates taken at the Royal Observatory, Cape of Good Hope under the direction of Sir David Gill, K.C.B., LL.D., D.Sc., F.R.SW., &c.'; *Annals of the Cape Observatory*, volume XII, part III, 1906.

DeVorkin, D.H. (2000). 'Internationalism, Kapteyn and the Dutch Pipeline', in: P.C. van der Kruit and K. van Berkel (editors): *The Legacy of J.C. Kapteyn, Studies on Kapteyn and the Development of Modern Astronomy*, Kluwer Academic Publishers, Dordrecht Boston London, 2000.

Forbes, G. (1916). *David Gill, Man and Astronomer*. See References of chapter 4.

Gill, D. and Kapteyn, J.C. (1896–1900). 'The Cape Photographic Durchmusterung for the Equinox 1875'; *Annals of the Cape Observatory*, III, IV and V, 1896–1900.

Guichelaar, J., Huitema, G.B., De Jong, H., editors, 2012. Articles by Guichelaar, J., Huitema, G. B. and Van Hoorn, M.; *Certainties in observations, Developments of the natural sciences in The Netherlands around 1900* (Dutch: *Zekerheden in waarnemingen, Natuurwetenschappelijke ontwikkelingen in Nederland rond 1900*), Verloren, Hilversum, 2012.

Hertzsprung-Kapteyn, H. (1928). *J.C. Kapteyn, His Life and Work* (Dutch: *J.C. Kapteyn, Zijn leven en Werken*), P. Noordhoff, Groningen, 1928.

Kapteyn, J.C. (1906). 'Star Streaming'; *Report of the Seventy-Fifth Meeting of the British Association for the Advancement of Science, South Africa, 1905*, John Murray, London, 1906.

Kapteyn, J.C., editor, (1908a). 'The proper motions of 3300 stars of different galactic latitudes, derived from photographic plates prepared by prof. Anders Donner, measured and discussed by prof. J.C. Kapteyn and Dr. W. de Sitter'; *Publications of the Astronomical Laboratory at Groningen*, 19, 1908. (Anders S. Donner (1854–1938) was a Finnish astronomer, who became good friends with Kapteyn during their work on the *Carte du Ciel*.).

Kapteyn, J.C., editor, (1908b). 'The parallaxes of 3,650 stars of different galactic latitudes, derived from photographic plates prepared by prof. Anders Donner, measured and discussed by prof. J. C. Kapteyn and Dr. W. de Sitter'; *Publications of the Astronomical Laboratory at Groningen*, 20, 1908.

Kapteyn, J.C. (1922). 'First Attempt at a Theory of the Arrangement and Motion of the Sidereal System'; *Astrophysical Journal*, 55, pp. 302–328, 1922.

Oort, J.H. (1935). 'In memoriam Willem de Sitter'; *The Observatory*, LVIII, 728, pp. 22–27, 1935.

Sears, F.H. (1922). 'J.C. Kapteyn'; *Publications of the Astronomical Society of the Pacific*, 34, 201, p. 233, 1922.

Smit, F. (2001). 'An outline of two centuries Royal Natural Sciences Society'; in: Blaauw, A. e.a. (editors), *A mirror of science*, Profiel Uitgeverij, Bedum, 2001. (Dutch: 'Een schets van twee eeuwen Koninklijk Natuurkundig Genootschap'; *Een spiegel der wetenschap*.).

Van der Kruit, P.C. (2015). *Jacobus Cornelius Kapteyn*. See References of Chapter 4.

Van Evert, J. (2012). *Oort, Researcher of the Milky Way and Founder of Radio Astronomy* (Dutch: *Oort, Melkwegonderzoeker en gondlegger van de radioastronomie*), Veen Magazines, Amsterdam, 2012.

Van Maanen, A. (1922). 'J.C. Kapteyn, 1851–1922'; *Astrophysical Journal*, 56, p. 145, 1922.

Warner, B. (1979). *Astronomers at the Royal Observatory Cape of Good Hope*. See References of chapter 4.

Chapter 7
Leiden Activities

> *Astronomy is more than any other science designated to be studied in its own right, to be pursued as an art.*
> De Sitter in his inaugural lecture (De Sitter 1908)

Inaugural Lectures of De Sitter and Van de Sande Bakhuyzen

In a combined meeting of the Board of Governors and the Senate[1] De Sitter took the oath as a professor on 21 October 1908. Directly after the oath he delivered his inaugural lecture: *The New Methods in the Mechanics of Celestial Bodies* [AUL, BG minutes] (De Sitter 1908). In a rough draft the title was *The New Methods in the Mathematical Theory of the Movements of the Planets* [ASL, Studies S8].[2] A few elements from his lecture are the following. The new developments in the study of the stars, the emancipation of the fixed stars, begun in the last quarter of the nineteenth century, had already achieved great successes, thanks to spectroscopy and photography. De Sitter made a connection to the evolution theory: the stars did not form anymore the fixed background, against which the study of the planets was performed and where proper motions and parallaxes actually were only annoying side effects; the groups of stars around us were part of a dynamical system, the evolution of which had become in the meantime part of astronomical research. De Sitter himself had been part for more than eight years of the headquarters of the new army, investigating the structure of the universe led by Kapteyn. The old astronomy of planets was considered inferior to the new star astronomy in certain circles. But, De Sitter argued, the astronomy of planets had undergone a rejuvenating cure.

The classical manner, starting from an undisturbed planetary orbit, to calculate repeatedly a next perturbation in order to acquire consecutively better approxima-

---

[1]The Senate (Dutch: Senaat) executed the daily management of the University and consisted of all professors, chaired by the *Rector Magnificus*, who was chosen yearly.
[2]For rough calculations and elaborations of ideas De Sitter used folio hardcover notebooks: numbered *Studies S1, S2, S3, ...*

© Springer Nature Switzerland AG 2018
J. Guichelaar, *Willem de Sitter*, Springer Biographies,
https://doi.org/10.1007/978-3-319-98337-0_7

tions, was one with mathematical problems. The expansions of the perturbation theory have been subpoenaed before the tribunal of the stern mathematics, and they have been questioned about their convergence. Then De Sitter examined the work of the American George W. Hill,[3] who had managed to determine the motion of the moon's perigee with an accuracy of 15 decimals, while the old theory was already faulty in the sixth decimal. Hill had not started with the Keplerian orbit of the moon around the earth in order to calculate the perturbations on it from the sun. No, he had chosen the values of earth, sun and moon slightly different from the real values, but by doing so he could solve the three-body problem exactly without any neglecting. Later the corrections had to be calculated and added, which was done by the English mathematician Ernest W. Brown.[4] The solution of the simplified three-body problem of Hill was periodic, leading to a far better possibility to predict future locations of the moon. Another important new mathematical development was the notion of *families of orbits*. These orbits were closely connected and the study of them and of the transitions between families gave a deeper insight into the development of a group of celestial bodies. Hill's new theory was mathematically based on Poincaré's work. De Sitter wished to develop his further investigations of Jupiter's satellites on the basis of Hill's and Poincaré's work. He said:

> The oldest of all sciences need not be inferior to her many centuries younger sisters in youthful vitality.

At last the appointment as an extraordinary professor of Van de Sande Bakhuyzen came through and he accepted his office on 2 June 1909 with the inaugural lecture *The Meaning the Older Measurements still have for the Astronomy Nowadays* (Van de Sande Bakhuyzen 1909). In his lecture Van de Sande Bakhuyzen discussed the proper motions of stars elaborately, for the first time shown by Edmond Halley[5] at the end of the seventeenth century. This motion was the key to the knowledge of our Galaxy, the effect we wish to determine. According to Van de Sande Bakhuyzen we patiently had to compare the measurements from different times, which had to be taken as distant from each other as possible, because proper motions were mostly very small. Although there was a lot of traditionalism in Van de Sande Bakhuyzen's views on research, there was nothing wrong with his opinion about the value of old observations in order to get a grip on the structure and dynamics of our Galaxy.

---

[3]The American mathematician and astronomer George W. Hill (1838–1914) worked mainly on three-body and four-body problems.

[4]The English mathematician and astronomer Ernest W. Brown (1866–1938) made extremely accurate tables of the moon. He worked mainly in the United States at Yale University and became an American citizen.

[5]The English astronomer Edmond Halley (1656–1742) discovered that the comets of 1531, 1607 and 1682 were one and the same, returning every 76 years. Now it is called Halley's Comet.

Fig. 7.1 George W. Hill

GEORGE WILLIAM HILL, PRESIDENT, 1895-1896

Education

When De Sitter came to Leiden the University had about 1500 students, of which just over 200 in the faculty of mathematics and the natural sciences. During his career as a professor until 1934 these numbers increased till roughly 3600 and 700 respectively. Mathematics and the natural sciences increased faster than the University as a whole. In 1908 there were 12 professors and two lecturers in mathematics and the natural sciences, and in 1934 there were 12 ordinary professors, 2 extraordinary professors,[6] 3 special professors[7] and 8 lecturers.

During his first years as a professor it took De Sitter a lot of time to prepare his lectures thoroughly. He lectured in the lecture hall of the Observatory. Lecturing in those initial years was not easy for him, as a number of comments in the Almanacs of the Leiden Students' Union from those years show. In the Almanac of 1910 we read:

> His Highly Learned[8] had to stop a series of lectures about the mechanics of celestial bodies at the request of the candidates,[9] because H.H.L. supposed too much mathematical knowledge from his audience. While this year was the first time H.H.L. had to lecture, we

---

[6]An extraordinary professor at a Dutch university is usually paid from a special university fund, and not from the regular educational budget.

[7]A special professor at a Dutch university has her/his main job mostly outside the university. The partial job at the university is often paid by an external organization.

[8]Dutch: Zijne Hooggeleerde or Z.H.G. It was a way of writing about a professor.

[9]A *candidate* was a student after having passed his candidate's exam (Dutch: kandidaatsexamen). Nowadays a student in the Master phase.

believe that gradually the presentation of the lectures will become more successful in the coming years.

The students were very fair to De Sitter. The same Almanac stated that the series of lectures on spherical astronomy of Van de Sande Bakhuyzen had unfortunately been minimally attended. The next year De Sitter lectured on the distribution of stars in the Galaxy (a subject of Kapteyn), but the writing down of many tables made his lectures a bit slow now and again. Van de Sande Bakhuyzen gave his lectures for only a very small auditorium, probably only one student. However, his demonstrations with the telescopes were fascinating. A year later De Sitter's lectures were already called *clear*. The Almanac of 1913 wrote about the introduction by De Sitter of a projection lantern, which provided important services. De Sitter's lectures had rapidly improved:

This year again extraordinary clarity and topical subjects were characteristic of the lectures.

In view of this rapid progress it is not surprising that later De Sitter was an advocate of lectures concerning pedagogy and didactics in the training of teachers. He had learned from the students' comments and had given much attention to the subject matter and the manner of teaching.

De Sitter also lectured on the method of least squares, with which calculations could be made on observational errors. After a few years, in 1916, this series of lectures stopped. De Sitter, together with De Jong, had written a small booklet on the subject (De Sitter and de Jong 1914), making a separate series of lectures unnecessary. The Almanac wrote that by the demanding presence at many lectures, one of the faults belonging to our academic life, there was no time left for individual study. The booklet of De Sitter was excellent and a recommendation for other professors. Also in the framework of better didactics De Sitter changed the form of his lectures to candidates in 1916. From then on they were given in the form of a colloquium. They were held primarily by students and led to fruitful exchanges of ideas.

The lectures of De Sitter were diverse: Fixed stars, Planetary system and motion of the moon, Mechanics of celestial bodies, Stellar astronomy, Observational errors, Least squares, Celestial mechanics and the expansion of the perturbation function, Principles of astronomy, Double stars and light curves, Determination of orbits and perturbations, Astrophysics, Spectral analysis, Recent discoveries and hypotheses in the field of stellar astronomy, Astronomical constants, and if need be also *Capita Selecta*. Van de Sande Bakhuyzen stayed mainly with his spherical astronomy.[10] In the archives of De Sitter's directorate a lecture notebook *Elementary Astronomy* of the student L. Binnendijk is present from the beginning of the 1930s, giving a broad survey of the astronomy, up to and including Oort's work on the spiral structure of the Galaxy.

In particular De Sitter had to prepare thoroughly his lectures outside his rather specialized field, like those about astrophysical subjects. Actually he considered his

---

[10]Almanacs of the Leiden Students' Union, in the University Library of Leiden.

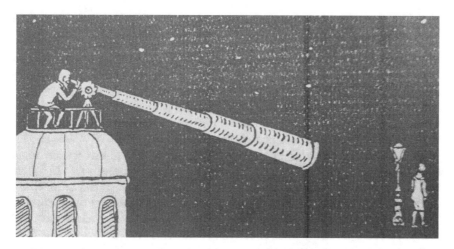

Fig. 7.2 De Sitter 'at work'. A cartoon from the Almanac of the Leiden Students' Union of 1934

own education too specialized. When he arrived in Leiden, there were no advanced students in astronomy. The enthusiasm to choose astronomy as major had already been declining under the old Van de Sande Bakhuyzen. In 1912 De Sitter had already five and their education required a lot of time [RGS, DOG/159, 3-1-1912 to Gill].

A few anecdotal reminiscences of the South African professor of astronomy Brian Warner[11] are worth mentioning. Warner was in the 1960s a colleague of Wesselink,[12] who was a student of De Sitter. Wesselink had followed lectures with De Sitter and had taken his doctoral degree with Ejnar Hertzsprung. Wesselink once told Warner that De Sitter set very complicated calculatory problems to his students. The answers of the students were all different from each other, and also different from those of De Sitter himself. But in the end he never told what the correct answers were. Wesselink's English was in the beginning not very good and when he showed De Sitter a text he had produced in English, De Sitter said: This is not good enough, do it again. But he gave no information at all how and where the text had to be improved. Perhaps De Sitter had forgotten that Mrs Kapteyn had to be his interpreter in 1896 during his first talk with Gill.

[11]The South African astronomer Brian Warner (1939) is emeritus professor at the University of Cape Town.

[12]The Dutch astronomer Adriaan Wesselink (1909–1995) worked in Leiden Observatory from 1929, where he wrote his dissertation under Hertzsprung, then deputy director. Later he worked in South Africa and the United States.

Relation with Van de Sande Bakhuyzen

The relation of De Sitter with Van de Sande Bakhuyzen was friendly, but that was about all that could be said. De Sitter had a room in the Observatory, where a computer worked for him on calculations related to the new theory of Jupiter's satellites [RGS, DOG/159, 25-4-1910 to Gill]. He gave his lectures in the lecture room of the Observatory. But they did not discuss the policies of the Observatory. Van de Sande Bakhuyzen had chosen to continue the policies of Kaiser and his brother. The modest Van de Sande Bakhuyzen did not try to involve De Sitter. Perhaps he was a bit frightened of the tall and self-assured aristocrat. The flamboyant De Sitter was his superior in social abilities, knowledge and intellect. De Sitter himself wished in no way to become involved in the Observatory policies. That was neither necessary in view of his investigations, which became more and more theoretical. Certainly when Albert Einstein and his general theory of relativity made their entrance into the Leiden scientific ambiance. He did all his work on the new theory of Jupiter's satellites nearly on his own. Later, the work on relativity and cosmology happened within the Leiden circle with Lorentz, Ehrenfest, Droste[13] en Fokker,[14] and by corresponding with Einstein, Eddington, Levi-Civita[15] and a few others. And if De Sitter needed contacts, measurements or photographic plates in the field of practical astronomy, he always had his contacts with South Africa and England.

In this way, de facto, a division in two had developed in the astronomy in Leiden, in spite of De Sitter's fine words directed to Van de Sande Bakhuyzen at the end of his inaugural lecture:

> With you, more than anybody else in this University, I will have to co-operate.

In his inaugural lecture nearly eight months later Van de Sande Bakhuyzen did not speak about cooperation. He kept to the word friendship. Cooperation between them could have had great advantages for students, staff and research at the Observatory. But that kind of cooperation was not given to De Sitter. His character was: either I am the boss, or I don't want to have anything to do with it. This character trait would again play a role a decade later in his sometimes difficult relation with his deputy director Hertzsprung.

---

[13]Johannes Droste (1886–1963) wrote his dissertation in 1916: *The gravitational field of one or more bodies, according to theory of Einstein* (Dutch: *Het zwaartekrachtsveld van een of meer lichamen volgens de theorie van Einstein*).

[14]Adriaan D. Fokker (1887–1972) was a student of Lorentz. Later he studied with Einstein and published on the theory of general relativity. He succeeded Lorentz at Teyler's Museum in Haarlem in 1928.

[15]The Italian mathematician Tullio Levi-Civita (1873–1941) made important contributions to the applications of the general theory of relativity.

Faculty Work

The professors of the faculty of mathematics and the natural sciences, of which astronomy was part, had a meeting roughly every month. The faculty meeting advised the Board of Governors. Important subjects were: proposing names of candidates for vacancies, giving advice on applications for scholarships, doctoral degrees and salary matters. The Minister also wished regular advice on matters like: the validity of foreign certificates, the approval to start a study with a certificate of the hbs, the cooperation and tuning of the fields of research between the corresponding faculties of the Dutch universities. Nearly everything passed the Board of Governors, which in the end made the decisions. De Sitter became secretary of the faculty meeting for the academic year 1910–1911. From every recently appointed professor it was expected to perform this task, so he had no choice. A few years later, he became dean of the faculty. In those years there was a lot going on concerning personnel of the faculty. Lorentz retired as an ordinary professor in 1912, after a career as professor of 34 years, and became an extraordinary professor until his retirement in 1923. He was appointed curator at Teyler's Foundation and Museum in Haarlem, in the province of North-Holland, where he could use the physical laboratory for his research. From there Lorentz stayed active and regularly gave his so-called Monday morning lectures in Leiden on actual developments in physics. Lorentz was of the opinion that Einstein would be his dream successor, but Einstein accepted a position in Zürich. Subsequently the faculty agreed to putting Paul Ehrenfest[16] first on the list of recommendations. He was appointed and started a major modernisation of the Leiden university education. Although Einstein did not come to Leiden as a professor, he was a regular guest there, because of his esteem for Lorentz and his friendship with Ehrenfest. From 1916 De Sitter would meet Einstein in Leiden regularly, when they discussed their ideas about cosmology.

De Sitter also took responsibility when ad hoc committees of advice had to be formed. In 1914 he formed, together with the pharmacist Leopold van Itallie (1866–1952) and the mathematician Pieter Zeeman (1850–1915), a faculty committee to write an advice on (the preparation) for higher education. Among other subjects, about the introduction of the so-called *lyceum*, a combination of the grammar school and the hbs, and about the question if an hbs certificate, without the classical languages Latin and Greek, should give access to a university education. Together with the physicist Johannes P. Kuenen (1866–1922) he wrote an advice on the alterations of the university library: they wished to enlarge the reading room. De Sitter was co-signatory of a proposal that was sent to the other universities for comment: universities had to support each other and pay less attention to their own

---

[16]The from origin Austrian physicist Paul Eherenfest (1880–1933) married the Russian Tatjana Afanasjeva (1876–1964) and they moved to St. Petersburg in 1907. Together they wrote the important *Begriffliche Grundlagen der statistischen Auffassung in der Mechanik*. His wife played an important role in the development of didactics of mathematics teaching in The Netherlands.

interests; each field of research and education had to be present on top level in one of the universities; difficulties had to be discussed with fellow scientists from other universities. As mentioned earlier, De Sitter also occupied himself with the pedagogical side of education.

> Mr. De Sitter considers it very important that teachers will learn more about the psychology and the methods of pedagogy. He considers it desirable that the faculty of mathematics and the natural sciences will be heard too before a possible appointment of a professor of pedagogy [AUL, Faculty minutes, 1916].

De Sitter's Health

Many years De Sitter's health was a limiting factor for his research and other activities. It is not quite clear what exactly the precise illness of De Sitter was that at times tormented him and kept him bedridden sometimes for longer periods. In view of his later cures it may have been a form of chronic tuberculosis. The recurring pattern of his illness seems to have started with a long illness in the summer of 1914, when he was confined to bed most of the time. In the beginning of September that year he had to have surgery. That surgery may also have played a role in the development of his later illness. He was glad that he could read and write again halfway October, he wrote to Hough in South Africa [RGO, 15/132, 12-10-1914]. De Sitter's old friend Schlesinger wrote in his obituary that De Sitter had an overdose of ether at a gallstone surgery, a result of which were his bad lungs (Schlesinger 1935). It remains uncertain if this was the surgery mentioned by De Sitter and the cause of his bad lungs.

De Sitter wrote to Frank Dyson[17] in 1916 that he was ill again, and that only God knew how long it would last [RGO, 8/88, 20-6-1916]. Anyway, De Sitter always had to take care of himself. In 1917 he was allowed leave of absence for a couple of months to stay in sanatorium Dennenoord in Doorn (province of Utrecht). At the end of 1918 Pannekoek wrote that he had heard the rumour that De Sitter was unwell [ASL, 11-12-1918]. De Sitter was hardly able to work, while the pressure to reorganize the Observatory was enormous. And at the beginning of 1919 Pannekoek was informed that De Sitter had decided to undergo a radical treatment. Pannekoek wrote to him that in that case Hertzsprung and he himself, the two intended deputy directors, had to be living at the Observatory to be able to play their executive role [ASL, 2-3-1919].

De Sitter had consulted his family doctor before accepting the directorship of the Observatory. The latter saw no important objections, he considered living at the

---

[17]The English astronomer Sir Frank W. Dyson (1868–1939) was Astronomer Royal of Scotland and later of England. He made Greenwich Mean Time more accurate and was one of the organizers of the expeditions to measure the sun's eclipse in 1919, which confirmed Einstein's general theory of relativity.

Observatory even an advantage, because he could reach his working quarters without going outside. At the end of 1919, when De Sitter had been busy with the reorganization for just a year, and when his deputy director Hertzsprung had only started recently, he went for a long cure to the Waldsanatorium in Arosa in Switzerland, which was specialized in treating tuberculosis. From there he wrote to Hertzsprung and other colleagues several times a week. But he mentioned his illness mostly in vague terms like: there is progress, or it will take at least so many months. Only after a stay of a year and a half De Sitter returned home at the beginning of April 1921.

One year later, after the first meeting of the International Astronomical Union in 1922 in Rome, where his wife accompanied him, De Sitter wrote to the president of the Board of Governors of Leiden University, that he was so tired that he asked for a few weeks of extra leave in order to recover in Florence. After that his health stayed reasonably well. When he made a tour through the United States for five months at the end of 1931 and the beginning of 1932, giving many lectures and visiting a lot of colleagues and observatories, his children were worried if it would not be too fatiguing for Adie (they called their parents Adie and Moedie). But after his long stay in Arosa the complaints were mainly gone. When his friend Schlesinger could only work at half power for a long time after a serious illness, De Sitter wrote that it had taken him years to become completely his own self again and had lost a couple of years productive work:

But that is all history now. Trust your doctor.

[ASL, 16-4-1931]

The Leiden Ambiance

One could presume that De Sitter must have felt a bit lonely in the University those first years. He had no institute with colleagues, like Kamerlingh Onnes. And he had little contact with the staff of the Observatory. But he had chosen for this position and it gave him, after the intensive work preparing his lectures, the possibility to do his research completely free, according to his own ideas, as he had once written to Gill.

Of course there were the contacts with the mathematicians and physicists in the faculty meetings and the contacts with colleagues from the other faculties in the meetings of the Senate. He felt at home in the circle of physicists in which Lorentz was the central figure.

Ehrenfest, the successor of Lorentz, asserted himself within the group of physicists. Lorentz came regularly from Haarlem to give his Monday morning lectures about actual developments in physics. In the period before Einstein published his final version of general theory of relativity (end of 1915), a group of theoreticians in Leiden was intensively busy with the new ideas. Einstein was very keen on contacts and correspondence with this group, consisting, as earlier mentioned, of Lorentz, Ehrenfest, De Sitter, Fokker and Droste (Kox 1988). They all published about Einstein's new theory in those years. Ehrenfest also admired De Sitter. By his admiration for others Ehrenfest often made unnecessarily less of himself. Much later, when De Sitter had been director of the Observatory for years, Ehrenfest wrote to his daughter:

> In our faculty there is only real activity around De Sitter, while he is (how awful!!) director of an observatory, while he is after all a THEORETICAL astronomer!

Ehrenfest always wrote with a lot of emphasis. Leiden was in those first years of the theory of general relativity the first scientific centre where the theory was accepted completely, from where is was also circulated.

A welcome addition to his scientific contacts was his appointment as a member of The Netherlands Academy of Sciences[18] in 1912. Later he would become a member of the Holland Society of Sciences.[19] The years from his appointment as a professor in 1908 until his appointment as director of the Observatory in 1918 would become scientifically extremely fruitful. The theoretical subjects covered a broad field. De Sitter always tried to tackle the fundamental issues. In the next chapters these theoretical investigations will be discussed.

The Netherlands Academy of Sciences

The old Van de Sande Bakhuyzen retired as active member from the Academy in 1912, of which he had been a member since 1872. By his retirement there was a vacancy for an astronomer. Willem de Sitter, then four years a professor in Leiden, was recommended with a hand-written document of more than four folio sheets, signed by the astronomers Kapteyn and the brothers Van de Sande Bakhuyzen, and also by Lorentz and Kamerlingh Onnes. The young Van de Sande Bakhuyzen had been a member since 1899. The document of recommendation contained a fine summary of De Sitter's work up to 1912. Point by point:

** Work: the satellites of Jupiter, discovery of the libration;
** Qualities: no small intellectual gifts, great perseverance, tough capacity to work, meticulous observation, sharp critic, alertness for systematic errors, command of the extremely difficult theory, proved by his solution of a special case of the four-body problem;
** Relativity: the meaning of relativity for the movements of celestial bodies under the influence of gravitation;

---

[18]Dutch: Koninklijke Nederlandse Akademie van Wetenschappen. In 1938 the Academy became Royal (Dutch: Koninklijk).

[19]Dutch: Hollandsche Maatschappij der Wetenschappen. In 2002 it became Royal.

** Practical astronomy: photometry, parallaxes (distance to the Hyades, 140 light-years, the until then largest determined distance), proper motions;
** Appointment: Foreign Associate of the Royal Astronomical Society in London in 1909.

With this recommendation and his already impressive list of publications it is not surprising that the department of mathematics and physics of the Academy put up De Sitter for election and chose him with 31 of the 41 votes. By letter of 15 May 1912 the Minister of the Interior informed the Academy of the royal assent for De Sitter's appointment. De Sitter thanked the Academy by letter of 18 May, writing that he was very sensitive for the honour bestowed on him.[20] With this membership De Sitter belonged to the scientific top of The Netherlands. The department of mathematics and physics convened every last Saturday of the month in the Trippenhuis[21] in Amsterdam. During these sessions the members had the opportunity to present scientific articles written by themselves, their students or non-member scientists, to appear in the regular *Reports* or in the *Proceedings* in the English language. During each meeting a few members gave a short lecture on a subject of their discipline. Besides there was of course ample opportunity to discuss scientific problems, candidates for vacancies, events at their universities or the writing of an advice. They could also discuss simpler questions. The highly learned gentlemen used to be annoyed during their sublime exchange of ideas by street rumour, caused by traffic on the cobble-stones of the street outside. They asked the city council to replace the pavement by wood or asphalt. A year later a request was sent to the Government for a subsidy for a new pavement. Whether the subsidy was granted could not be found in the archives.

The Academy as (Inter)National Advisory Board

The Academy was an important advisory board of the government concerning scientific matters and it was asked for counsel concerning imminent bills. An example was the positive advice that was given about the parliamentary bill of the liberal democrat Joseph Limburg (1866–1940). In the bill was proposed that students with an hbs exam could enter university in the faculties of mathematics and physics, and medicine. First the bill was presented because it became more and more clear that a scientific study in these faculties could be brought to a successful conclusion without an additional exam in the classical languages, and secondly because young boys lost considerable time already by the mobilization as a result of the First World War. The department mathematics and physics wrote a clear letter

---

[20]The famous mathematician Luitzen E. J. Brouwer (1881–1966) became a member on the same date.
[21]The Trippenhuis had been built in 1660 by the brothers L. and H. Trip, arms merchants. Since 1812 it is the official seat of the Academy.

Fig. 7.3 Willem de Sitter
painted by the Dutch painter
Jan Sluijters in 1934. The
picture is in the possession of
Leiden observatory. The
photograph is taken by
Ph. E. Habing

to the Minister of the Interior on 16 March 1917, stating that already too many had been obliged to make considerable efforts and had lost precious time. The bill became the Limburg law.

The actual political situation was also a subject of discussion. Immediately after the outbreak of the First World War chairman Lorentz spoke about the gravity of the time and in 1915 he proposed, with a view of the circumstances not to appoint foreign members. That policy was continued for a number of years. On 30 November, immediately after the end of the First World War, Lorentz pleaded for a swift restoration of the international cooperation. The mathematician Diederik J. Korteweg (1848–1941) said that Lorentz was the ideal man to reinstate this cooperation. In the period following the First World War many discussions in the Academy meetings were devoted to the new international scientific cooperation. The *International Research Council* (French: *Conseil International des Recherches*) was founded, in which the former central powers (Germany, Austria–Hungary, Ottoman Empire, Bulgaria and its allies) were not yet allowed to take part. The Academy set up The *Scientific International Cooperation Committee*[22] in 1923, in order to outline a new course. De Sitter was a member of this Committee, representing the astronomers.

---

[22]Dutch: Wetenschappelijke Internationale Samenwerkingscommissie (W.I.S.-Commissie).

Internal Academy Committees

Much later, in 1932, the Academy addressed a request to the Minister to abolish the *Amsterdam Time*. The request was based on a report drawn up by De Sitter et al. Internationally it had been decided in 1884 to define 24 time zones. But The Netherlands had decided not to join. Even in 1907 The Netherlands did not yet manage to decide: Western European Time or Central European Time.[23] Up to then the time varied from place to place in The Netherlands. After long discussions a half-baked decision was taken in 1909 to choose the Amsterdam Time for the whole of The Netherlands: Greenwich Mean Time + 19 min and 32.13 s. In fact it was the exact time of the meridian going through the tower of the Western Church[24] in Amsterdam. But eventually The Netherlands stood alone in Europe. Especially in the sciences there was regularly lack of clarity. Astronomers always had to make extra corrections in order to be able to communicate with their foreign colleagues. In a letter to the Minister the Academy stated:

> The Department of Physics of the Academy of Sciences urges the government to free The Netherlands as soon as possible from their isolation, by abolishing the Amsterdam Time and adopting the World Time.
>
> [NAW1]

The request was not successful. Not until 1937 a small change was implemented: the time in The Netherlands became Greenwich Mean Time + 20 min. An incredible improvement of 27.87 s. One day after the capitulation to the Germans in the Second World War, the Central European Time was introduced, which stayed the official time after the war.

De Sitter was chairman of the Eclipse Committee of the Academy, although he never joined an expedition to a solar eclipse, perhaps in view of his weak health (with the exception of the circular, and not complete, eclipse in Maastricht in the province of Limburg in 1912). The Eclipse Committee was founded in 1899, with H. G. Van de Sande Bakhuyzen as chairman and the recently appointed professor from Utrecht Nijland as secretary. At every eclipse they tried to set up an expedition, starting in 1901 on Sumatra in the Dutch East Indies. An official assignment from the Dutch government and considerable subsidies, also by private individuals, were necessary for each expedition. In view of the focus on the sun of much investigations in Utrecht, in particular under the direction of professor of physics Willem H. Julius (1860–1925), the Utrecht Observatory always played an important role in the organization of the eclipse expeditions. An eclipse expedition had to be prepared meticulously, because of the extremely short duration of the complete eclipse. Only then a maximum of observations could be performed in those few

---

[23]Western European Time is equal to the Coordinated Universal Time or UTC (and to Greenwich Mean Time or GMT). Central European Time is 1 h ahead of the UTC (=UTC + 1).
[24]Dutch: Westerkerk.

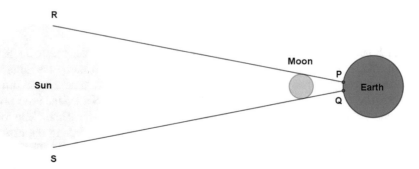

Fig. 7.4 Solar eclipse

minutes. An important aim was always the investigation of the sun's corona, which otherwise could never be seen (Van Berkel 2004).

Solar eclipse
When the moon is between the sun and the earth, there is a cone-shaped shadow on the side of the earth, from where the sun cannot be seen. If the moon in his ellipse orbit is slightly more near the earth, the apex of the cone is below the surface of the earth. Then there is an area on the earth from where the sun cannot be seen: a total eclipse. If the apex of the shadow cone does not reach the earth by a slightly greater distance of the moon, then there is an area on the earth where the central part of the sun's disc cannot be seen, while the outer rim may still be seen. In Maastricht there was such a ring-shaped eclipse in 1912.

As a result of the rotation of the earth round its axis alone the small area of a total eclipse would move on the surface of the earth from east to west. As result of the rotation of the moon round the earth alone the area would move from west to east. The second effect is greater than the first and the area of total eclipse moves thus in an easterly direction. On a single point on the earth's surface the shadow cone passes in a few minutes. On the most advantageous point, in the centre of the path of passage it can be five or six minutes.

In the figure we see the conic space between the extreme light rays RP and SQ. Inside this cone the sun is completely hidden behind the moon. In the area PQ on the surface of the earth there is a complete solar eclipse.

The Holland Society of Sciences

De Sitter became a member of the old Holland Society of Sciences in Haarlem in 1922. This Society was founded in 1752 with the aim of promoting the sciences. Then Lorentz was secretary of this oldest learned society of The Netherlands. It gave De Sitter a new set of contacts and possibilities. The Society was a great supporter of astronomy in those years. The Society had decided positively on a request for a subsidy by De Sitter and Nijland. It had awarded a sum of 200 Dutch guilders yearly for a period of five years to make possible the edition of the *Bulletin of the Astronomical Institutes of the Netherlands*. And the Pieter Langerhuizen Lambertuszoon Foundation, of which The Society was in charge, awarded a subsidy of 4500 guilders to set up a completely new measuring programme to free declination measurements of stars from systematic errors. In the chapter on International Projects this measuring programme is discussed [HMW].

References

De Sitter, W. (1908). *The New Methods in the Mechanics of Celestial Bodies* (Dutch: *De Nieuwe Methoden in de Mechanica der Hemellichamen*), inaugural lecture, Gebroeders Hoitsema, Groningen, 1908.

De Sitter, W., C. de Jong, editor (1914). *Introduction to the Method of the Least Squares, according to the lectures given at the University of Leiden by Dr. W. de Sitter, at the request of the Philosophical Faculty of the Leiden Students* (Dutch: *Inleiding tot de Methode der Kleinste Kwadraten* etc.), Boekhandel en Drukkerij, voorheen E.J. Brill, Leiden, 1914.

Kox, A.J. (1988). 'General Relativity in the Netherlands 1915–1920'; in: Eisenstaedt, J. and Kox, A.J. (editors), *Studies in the History of General Relativity*, pp. 39–56, 1988.

Schlesinger, F. (1935). 'The Progress of Science'; *Scientific Monthly*, 40, pp. 89–90, 1935.

Van Berkel, K. (2004). 'The Academy and the Second Golden Age' (Dutch: 'De Akademie en de Tweede Gouden Eeuw'); *KNAW*, Amsterdam, 2004.

Van de Sande Bakhuyzen, E.F. (1909). *The Meaning the Older Measurements still have for the Astronomy Nowadays* (Dutch: *De Beteekenis die de Oudere Waarnemingen nog Heden voor de Sterrenkunde hebben*), inaugural lecture, Boekhandel en Drukkerij voorheen E.H. Brill, Leiden, 1909.

Chapter 8
Fundamental Work and a New Theory of Jupiter's Satellites

More and more I find myself on the path of pure theoretical celestial mechanics.
De Sitter in a letter to Gill [RGO, 15/130, 27-3-1902]

Newton's Gravitation and Perturbation Theory

De Sitter was obsessed by gravitation during his whole working life. Till Einstein there was only Newton's law of gravitation, giving the gravitational force F_g of two masses m_1 and m_2 on each other: $F_g = Gm_1m_2/r^2$, in which r is the mutual distance and G the gravitational constant (a mass twice as big gives twice the force, a double distance gives a quarter of the force). With the help of the tools provided by the celestial mechanics De Sitter worked on the heliometer measurements of Gill, in order to deduce from them data of Jupiter's moons. To this purpose he made use of perturbation theory in the celestial mechanics (see Chap. 5). For the calculation of the orbit of a single moon around Jupiter it was simple: the orbit is an ellipse. But the other moons, although having a relatively small mass, but not far away, and also the sun, although far away, but with a large mass, influenced by their gravitational force the orbit of the first moon and so caused small perturbations in the perfect ellipse. These perturbations can change for instance the eccentricity or the direction of the major axis of the ellipse. These small changes may be periodical (always returning after a certain period) or secular (continually increasing or decreasing). It is an extremely complicated mathematical calculation to deal with all these perturbations. The influence of the other planets (relatively small masses as well as large distances) may fortunately be neglected, because these perturbations are orders of magnitude smaller. Always it had to be decided upon which perturbations had to be taken into account and which not. It was also necessary to include data, calculated by colleague-astronomers. But De Sitter loved such a complicated mathematical theory. Mathematics was after all his first love. Indefatigably he enjoyed the long and tedious bouts of calculation. Numerous notebooks, called *Studies*, numbered *S1, S2* etc. in the archive bear witness to this. A related aim of all

© Springer Nature Switzerland AG 2018
J. Guichelaar, *Willem de Sitter*, Springer Biographies,
https://doi.org/10.1007/978-3-319-98337-0_8

Fig. 8.1 The family in their back garden in Leiden around 1910. Note the hammer and cigar in De Sitter's right hand. The photograph is from the family archive of De Sitter's granddaughter Tjada van den Eelaart-de Sitter

this work was to produce better *ephemerides*. These lists with positions in the sky of heavenly bodies for future times had been published for Jupiter and its satellites up to the nineteenth century. But at the end of this century the prevailing ephemerides turned out to be increasingly less accurate. The differences between the observations and the predictions became larger all the time. For De Sitter an extra reason to intensify his studies into Jupiter's moons.

Aspects of De Sitter's Fundamental Work

Shortly after his doctoral degree he had dived also into the purely theoretical celestial mechanics. And later, in his inaugural lecture in 1908, he laid down the basics of his programme for further investigation. But: in all his theoretical work De Sitter made the connection with experimental results. De Sitter would expand the method used by Hill for the solution of the three-body problem, to a simplified four-body problem of Jupiter and its three inner moons, which he could solve completely. He used for this solution an intermediate orbit. After that he would calculate the perturbations on that solution caused by the flattening of Jupiter, the fourth moon and the sun. This method would in the end give better results for the Jovian system. De Sitter did not yet mention his plans concerning Jupiter's satellites in his inaugural lecture explicitly, but Hill's method kept running through his mind and he would explain his plans to Gill shortly afterwards. He already published about it in 1909 and he would finish the new theory ten years later. He compared his theoretical results to observations made over a period of more than two centuries. Besides celestial mechanics in his inaugural lecture De Sitter had discussed the origins of the law of gravitation as well, although this actually lay outside his main subject:

> The law of gravitation has not yet been affected, it has not yet found a satisfying explanation.

> Gravitation is not subject to absorption, not to refraction, no speed of propagation has been established, it works equally on all bodies indistinguishably, everywhere and always we find it in the same strict and simple form, on which all our efforts to penetrate into its internal mechanism rebound.

Still this elaboration was inserted deliberately. The deeper meaning and origin of gravity, the elusiveness of this fundamental force, on which all his work was based, kept his mind occupied as a matter of course. Later, after accurate study of measurements, he consigned to the waste bin the supposition that the gravity between two masses could be absorbed by masses in between. The necessity to penetrate into the mechanism of gravity was enhanced by the not yet solved problem of the unexplained extra perihelion shift of the planet Mercury. Mercury, the planet nearest to the sun, has always occupied the astronomers, especially due to its special orbital elements (nearest to the sun and by far the largest orbital eccentricity). If we assume that Mercury moves round the sun under the sole influence of the sun's gravity, the planet will describe a fixed ellipse, with the sun in one of its foci (the sun's mass is extremely larger than that of Mercury). But as a result of small disturbances, for instance gravitational pulls by other planets, the form of the ellipse stays nearly unaltered, but the long axis of the ellipse rotates slowly. This perihelion shift was known accurately, after centuries of measurements: 574″ (seconds of arc) per century. The French astronomer Urbain Le Verrier (1811–1877) succeeded in calculating the shift due to the influence of the other planets basing himself on the then known theory. His result was: 532″ per century. But the remaining 42″ per century stayed a mystery. Le Verrier suggested as a solution of the problem the existence of another planet, rotating round the sun inside Mercury's orbit. He called the new planet Vulcan. However, later calculations proved that Vulcan had to be visible due to its size. It was never observed and in the end the search for Vulcan was ceased.

A different possible explanation was given by the German astronomer Hugo von Seeliger (1849–1924). He supposed that large clouds of dust were rotating round the sun inside Mercury's orbit. These clouds of dust could then explain the extra perihelion shift. But the dust had in certain circumstances to be visible by the scattering of sunlight, the zodiacal light. This phenomenon could in the end not be observed either. Other explanations were proposed: the oblateness of the sun, or a

Fig. 8.2 Urbain Le Verrier

slightly different exponent in Newton's law of gravity ($F_g = \gamma m_1 m_2 / r^{2 \pm \varepsilon}$). But neither of these hypotheses held up after accurate calculations and measurements. The final solution was given by Einstein at the end of 1915, on the basis of his general theory of relativity.

Ten years before the publication of the general theory of relativity, Einstein's special theory of relativity had been published in 1905. The word *special* meant that only co-ordinate systems moving relative to each other at constant velocity were considered. In this theory the basic concepts of celestial mechanics: space, time and speed of light, were connected to each other in a fundamental new way. Once De Sitter, with his interest in fundamental concepts, was appointed in Leiden, it was very understandable that he studied the new theory thoroughly, in search of the consequences on his work in celestial mechanics. It resulted in a long article: 'On the Bearing of the Principle of Relativity on Gravitational Astronomy' (De Sitter 1911). Later he underpinned Einstein's supposition of the constant speed of light with observations made of double stars. And after the publication of the general theory of relativity in 1915 De Sitter, with his ample mathematical gifts, would dive into the new theory, which at last penetrated into the internal mechanism of gravitation. It would lead to the publication of his three famous articles in 1916 and 1917, published in the *Monthly Notices*, which would make known the general theory of relativity in England. In the third article in this series De Sitter published the cosmological model named after him (see Chap. 9). Already in this article De Sitter tried to make a connection with the first observations of receding galaxies. To finish these aspects of De Sitter's fundamental works his interest in astronomical constants must be mentioned. He was aware of the importance of these constants. On the basis of all data known to him, he published twice on these constants. The second time his student Dirk Brouwer processed De Sitter's data and wrote a posthumous publication.

Fig. 8.3 Hugo von Seeliger

De Sitter's Publications

De Sitter's publications contain two main groups: the articles on celestial mechanics and those on relativity and gravitation: 64 and 70 respectively.

The data for the following survey of articles are mainly drawn from the dissertation of Röhle (2007).

The publications of De Sitter are conveniently classified in four periods:
** his training period with Gill and Kapteyn (1899–1908);
** the productive period as a professor before his appointment as director of the Observatory (1909–1919);
** the managerial period: director of the Observatory, work in the Senate of the University, secretary and president of the International Astronomical Union (1920–1929);
** last years with emphasis on the expanding universe (1930–1934).

Fields of research	Periods			
	1899–1908	1909–1919	1920–1929	1930–1934
Celestial mechanics, Jupiter's satellites	21	22	21	1
Sun, earth, moon, planets, astronomical constants, time	1	11	17	1 (1938)

(continued)

(continued)

Fields of research	Periods			
	1899–1908	1909–1919	1920–1929	1930–1934
Gravitation and relativity	0	30	10	30
Stars	8	2	2	0
Measuring apparatuses	3	0	2	0
Various	1	2	23	16
Subtotals	34	67	75	48
Total: 224 articles				

A few points catch the eye:

** The constant series over a period of thirty years of publications on Jupiter and its satellites. In 1930 De Sitter stops with his Jupiter research. A total of 64 articles.

** In the second period his articles discuss the special as well as the general theory of relativity. A total of 30 articles.

** In De Sitter's managerial period this number decreases to 10.

** In the fourth period the articles mainly discuss the expansion and the structure of the universe. A total of 30 articles.

** The articles on stars date mainly from his period with Kapteyn.

** The various articles are mainly lectures and yearly reports as director of the Observatory.

** The article from 1938 is the posthumous article edited by Brouwer.

In the following paragraphs of this chapter and in Chap. 9 De Sitter's most productive scientific period is discussed.

In the remaining paragraphs of this chapter first a number of articles by De Sitter on the special theory of relativity and gravitation are treated, to conclude with a discussion of his new theory for the four Galilean satellites of Jupiter.

In Chap. 9 his survey articles on the general theory of relativity, his new cosmological model and his discussions with Einstein are treated. His own solution of the Einstein equations became known as the De Sitter Universe.

The special theory of relativity
Relativity principle and constant speed of light
De Sitter not only wished to get to the bottom of a theory but also calculate its consequences for real astronomical problems. The special theory of relativity, published by Einstein in 1905, was based on two principles.

First: all physical laws must have the same form, be covariant, in each system of co-ordinates. But: the theory only deals with co-ordinate systems moving with constant velocity relative to each other.

Second: the speed of a light ray is the same for every observer, independent of the his co-ordinate system. The speed is nearly: 300,000 km/s. This second principle is not true in the old Newtonian mechanics.

Lorentz transformations

The two principles together lead to the so-called Lorentz transformations between two co-ordinate systems. The principle of the constancy of the speed of light more or less mixes the notions of space and time. Lorentz had derived transformations between the co-ordinates of two systems moving relative to each other in order to solve the problem of the negative results of the Michelson-Morley experiments without giving up the ether as medium for light waves.

But Einstein said: if we cannot observe and measure the ether, we might as well skip it completely. With his principles Einstein could derive the transformation formulae too.

The co-ordinates of a single event in co-ordinate system S(x, y, z, t) and those of the same event in system S'(x', y', z', t') are connected by the following formulae. The system S' moves with a constant velocity v in the positive x-direction with respect to S.

First we give, for comparison the formulae in the old Newtonian theory:

$$x' = x - vt, y' = y, z' = z, t' = t.$$

Second the relativistic formulae, in which we use the abbreviation

$$\gamma = 1/\sqrt{(1 - v^2/c^2)} :$$

$$x' = \gamma(x - vt), \ y' = y, \ z' = z, \ ct' = \gamma(ct - (v/c)x).$$

Two peculiar effects are time dilation and length contraction.

** Let T be a lapse of time measured in a co-ordinate system S by an observer at rest in S. This lapse of time measured by a different observer at rest in co-ordinate system S', moving at constant velocity v relative to S is T'. We have the T' in formula: $T' = \gamma T$, a time dilation.

** Let a rigid rod with length L' fixed in co-ordinate system S', be moving in the direction of the rod with a velocity v, relative to co-ordinate system S. When the rod is measured by an observer in S, he finds length L, with: $L = (1/\gamma)L'$, a length contraction.

The Influence of the Special Theory of Relativity on Gravitational Astronomy

De Sitter had made a thorough study of Einstein's special theory of relativity of 1905 and its application on astronomy. His appointment as an Associate of the Royal Astronomical Society made it easy for him to publish in their *Monthly Notices*. At the end of 1910 he wrote to Gill that he had reached a number of interesting results, among others an extra perihelion shift of Mercury, comparable to Von Seeliger's result with his idea of dust clouds [RGS, DOG/159, 24-12-1910]. In March 1911 'On the Bearing of the Principle of Relativity on Gravitational Astronomy' appeared. De Sitter wrote that there was a perihelion shift according to the principle of relativity, if Mercury orbited round the sun without perturbations of other planets. That could also have its influence on the periodic solutions of the three- and four-body problems in celestial mechanics. From a theoretical point of view that was of course relevant for his new theory for the satellites of Jupiter. But he also wrote the article as a practical astronomer. He tried to answer the question: what are the effects of the principle of relativity which can be measured? He hardly made reference to literature. He did not mention Einstein, he did mention Poincaré, Minkowski[1] and Lorentz. He thanked Lorentz for the many conversations on the subject. De Sitter set himself two tasks in the article:

** Which gravitational force law comes in the place of the classical Newtonian gravitational law?
** How can we rewrite the movements in one co-ordinate system in another one?

He then developed the formulary necessary for his calculations himself, because he found that easier than to dig into the literature. Moreover, he supposed that the gravitational influence of one body onto another propagated with the speed of light, although this does not follow from the relativity principle. In order to derive the force law complying with the relativity principle he made use of Minkowski's work. After all he made approximations by a number of series expansions. For the perihelion shifts of the small planets he found the values given in the table.

Perihelion shifts in seconds of arc per century	
Mercury	7.15
Venus	1.43
Earth	0.639
Mars	0.225

---

[1]The German mathematician and theoretical physicist Hermann Minkowski (1864–1909) developed a description of the special theory of relativity, in which time and space co-ordinates were treated on an equal footing: the four-dimensional Minkowski spacetime.

For Mercury it was 7.15″ per century, by far not the missing 42″ per century.
Next De Sitter applied this technique to the orbit of the moon and found a perigee
shift of 1.172″ per century. Unfortunately too small to be measured.

Another result was also completely immeasurable: the decreasing shift of the
vernal point of 0.0000044″ per century. By the precession of the axis of the earth
the vernal point makes a complete rotation in 26,000 years, which is nearly 5000″
per century. Further De Sitter focused again on a few fundamental points, for
instance the question about the relation of the astronomical time, determined by the
rotation of the earth, and the variable t used in the several co-ordinate systems he
used. Although the results of this article were limited as measurable astronomical
quantities were concerned, De Sitter was one of the first astronomers investigating
systematically the consequences of the relativity principle. With this work he set a
new step to the solution of the mysterious excess of Mercury's perihelion shift. He
did that already in a period in which there were still doubts about the special theory
of relativity, even in scientific circles.

Absorption of Gravitation

Already in his inaugural lecture in 1908 it had emerged how the essence of gravity
occupied him. Would the gravity of the sun on the moon during an eclipse of the
moon[2] become smaller as a result of the fact that the earth is *in its way*? In his
inaugural lecture De Sitter had mentioned this possible effect. He was probably the
first to have done so. Von Seeliger wrote about it in 1909. He considered the
absorption effect plausible and measurable. In December 1909 and January 1910
De Sitter made calculations without knowledge of Von Seeliger's publication. He
wrote to Gill that he wished to send quickly an article on the subject for publication
in the *Monthly Notices* [RGS, DOG/159, 25-4-1910]. In that article he wished to
rate the absorption by the earth at 1/300,000 part of the gravitational force of the
sun. In that way he could perhaps explain a large part of the fluctuations in the
length of the moon (the measured direction of the moon relative to the calculated
direction without absorption), although he did not characterize these ideas as the
real origin of the phenomenon of the fluctuations, but more like a suggestion, worth
to be published. These fluctuations had been calculated by the Canadian–American
astronomer Newcomb[3] from old observations, as far back as 1672, and they could
not be explained on the basis of the regular theories.

---

[2]During an eclipse of the moon the earth is in between the moon and the sun. The moon passes
through the conically shaped shadow behind the earth.

[3]Besides by astronomy Simon Newcomb (1835–1909) became known by a variety of subjects. He
worked together with Michelson to measure the speed of light. He was the first to discover
Benford's law (in a much used table of logarithms the pages starting with a 1 are used more than
the other pages). He doubted very much the possibility of a flying machine.

Fig. 8.4 Simon Newcomb

However, in May De Sitter wrote to Gill that the article was not yet ready to send in, he had found an error in the calculations [RSG, DOG/159, 4-5-1910]. Some time elapsed, during which De Sitter could not work on it. To achieve a good result, a much more elaborate calculation had to be performed. At the end of 1911 De Sitter learned that a young German, Kurt F. Bottlinger (1888–1934), a student of Seeliger, worked on the problem and he decided to await his results. A short article of Bottlinger appeared in the *Astronomische Nachrichten* in January 1912, in which he reached a good agreement with Newcomb's results, on the basis of an absorption of 1/60,000 (Bottlinger 1912). He processed his results in a dissertation later that year. Then De Sitter took up the subject again. A short article in *The Observatory* appeared in November 1912, in which he compared the results of Newcomb, Bottlinger and his own preliminary results, making use also of results published by Ernst van de Sande Bakhuyzen in 1911 (De Sitter 1912). There was good agreement between measured and calculated fluctuations with a period of roughly 18 years from 1830 until 1900, but not in the direction of the line on which the fluctuations were superposed. De Sitter concluded that the case was not yet closed. There were two conclusions possible: absorption was not the correct explanation, or the calculations were not accurate enough. As the only way out he saw a much more accurate calculation over a period of a couple of centuries, which also should explain the general directon. In the last sentence he congratulated Bottlinger with his excellent work.

De Sitter was not quite sure and executed his plan to calculate everything again, but with much greater precision. In 1913 he presented a communication to the Academy of Sciences in the form of two articles of 16 and 17 pages (De Sitter 1913a). In the English language they would also appear in the *Proceedings of the Academy*. In these articles, De Sitter supposed that the absorption was proportional with that part of the mass of the earth between the sun and the moon. The density distribution within the earth was also something to reckon with. Just like Bottlinger

De Sitter took the data from the German seismologist J. Emil Wiechert (1861–1928) for this density distribution: a core of 0.77 times the radius of the earth with a density of 8.25 g/cm^3 and a mantle with a density of 3.30 g/cm^3.

The comparison of his results with the observations led him to a few conclusions. In a number of shorter periods during the past centuries the similarities were reasonable, which had brought Bottlinger to the conclusion to have shown the absorption with a good probability. But in the rest of the periods there remained, after subtraction of the results of the calculations, discrepancies in the length of the moon. These unexplained differences had to have according to De Sitter a different cause, but less irregular than the original. However, the residues showed an even more irregular course. That was not possible, in the opinion of De Sitter. De Sitter finished part one with the conclusion that for the time being there was no reason to accept the existence of a measurable absorption of gravity as proven, or even probable. In the second communication De Sitter repeated his calculations with a different density distribution of the earth, yielding the same results.

He had thought much about it, he had calculated a lot, perhaps he had hoped to be able to prove absorption of gravity, but in the end he drew the only scientifically correct conclusion: there was no absorption of gravity.

Ether or no Ether

When in the nineteenth century the wave character of light had been shown, the necessity was felt for the existence of a medium, the ether, in which the light waves could propagate. Lorentz was convinced of the existence of an ether at rest. The earth, moving with a speed of 30 km/s round the sun, and thus through the ether, had to feel an *ether wind*. By the varying direction of the motion of the earth in its orbit round the sun the measured speed of light had to change. However, the experiment of Michelson and Morley,[4] in which they measured the speed of light in two perpendicular directions, showed no differences in the measured speeds. Their results played an important role in abandoning the concept of the ether. Lorentz proposed two hypotheses to break the deadlock: a slightly different force between the molecules and a slightly different form of the electrons moving through the ether. These hypotheses led to the so-called Lorentz transformations giving the relations between the co-ordinates of two co-ordinate systems moving with a constant velocity relative to each other. With these transformations Lorentz could prove that no ether wind could be measured. Einstein cut the Gordian knot in his own manner. If you can in no way measure the ether after all, let us then assume that it does not exist. With his relativity principle and the supposition that the speed

---

[4]The Prussian-American physicist Albert A. Michelson (1852–1931) and the American physicist Edward W. Morley (1838–1923) performed their experiments in 1887.

of light is equal for every observer he could explain the results of Michelson and Morley and derive the Lorentz transformations. For many physicists the abandoning of the ether was a big step: the light waves had to have *something* to wave in?

But there came an alternative theory, the emission theory of the Swiss theoretical physicist Walter Ritz (1878–1909), who had written a critical paper about the theory of Einstein and Lorentz in 1908. The essence was that light is emitted from its source with a speed c relative to that source. He supposed that, if an object is moving to an observer (with a velocity v relative to the observer) and is emitting a light ray in the same direction (with a speed c relative to the source), this light signal will move to the observer with a speed c + v. So also without the ether there were two theories: an *experimentum crucis* was necessary.

De Sitter Provides an *Experimentum Crucis*

Paul Ehrenfest, appointed as successor of Lorentz, delivered his inaugural lecture on 4 December 1912. The subject was the crisis of the ether hypothesis.

De Sitter was in the audience and enjoyed himself. Here, someone was lecturing, who would bring a wave of reform in education and research. De Sitter later wrote in an obituary after the suicide of Ehrenfest (De Sitter 1933):

> There was nothing professorial on him. The fact that he spoke from the podium and not behind a lectern, gave him the opportunity to display his exceptional vitality and unusual agility. This made the images he used visible and palpable for us, also enhanced by his rich metaphors.

Ehrenfest must have impressed De Sitter, remembering the details so vividly after more than twenty years. One of the elements in his lecture was a discussion about an *experimentum crucis*, a deciding experiment to choose between Einstein's theory and the one by Ritz: Is the observed speed of light equal for every observer, or does it also depend on the velocity of the emitting source?

The idea of Ehrenfest was surprisingly simple: take two light sources, one at rest relative to an observer and a second one moving fast in the direction of the observer. They both emit a light signal and the observer measures if they pass through a tube in front of his eyes in the same time or not. However, on earth this experiment could not be done with the required accuracy.

De Sitter understood that the astronomy could possibly supply this experiment by looking at double stars (binary double stars, orbiting round their common centre of gravity). It is not known if the idea immediately occurred to him still listening to Ehrenfest, and that he hurried home after the congratulations to start double star calculations. See for De Sitter's idea the following explanation.

Einstein versus Ritz

To follow De Sitter's idea, take a double star at distance d from the earth, with the line of sight from the earth in the plane of rotation. Take for simplicity one star with a large mass M and a second star with a small mass m. The small mass makes a circular orbit with radius r round the large mass with velocity v.

Einstein

Taking into account the constancy of the speed of light c, all light rays from each star reach the earth in the same time. Let us neglect the time differences because of the varying distance from the small star to the earth. We see the large star staying in one direction and the small star going from left to right and back in the same time. Another observer (on a planet half way) would see exactly the same.

Ritz

In the case of Ritz the star sends from its extreme points left and right light rays with velocities $c + v$ (star moving to the earth) and $c - v$ (moving from the earth) respectively. They do not reach the earth at the same time. The time of the light ray from the receding star is $d/(c - v)$ and the time of the light ray of the approaching star is $d/(c + v)$. This leads to an asymmetry in reaching the extreme points. The time from left to right is longer than the time from right to left.

Here De Sitter had his *experimentum crucis*. From all available observations on double stars an asymmetry was not to be found. So the principle of Ritz could not be true and the speed of light for every observer had to be the same constant value. De Sitter had quickly elaborated his idea, as can be seen in his notebook *Studies S2* [ASL]. In February 1913 he already offered a short article on the subject to the *Physikalische Zeitschrift* (De Sitter 1913b). During the meeting of the Academy of Sciences on Saturday 22 February he offered a slightly differently formulated article for the *Reports of Physics* and besides the same one in English for the *Proceedings*. Both these articles ended with the remark that, even starting from Ritz' principle, the light signals measured on earth could very well be in agreement with a proper Keplerian orbit, but then the stars had to traverse one half of their orbit much faster than the second half, in order to compensate for the time differences of the light rays on their way to the earth. De Sitter used the qualification *absurd*. This word appears more or less regularly in writings of De Sitter. He loved to express himself sometimes with emphasis. In this case he could not refrain from a bit of sarcasm.

This conclusion of De Sitter was an important result, which did not escape Einstein's attention. After all the assumption of the constancy of the speed of light was a fundamental hypothesis in his theory. The experimental confirmation of it brought his hypothesis on the higher level of a verified scientific fact.

To the New Theory of the Jovian Satellites

Poincaré and Hill
Soon after the publication of his dissertation De Sitter had entered the field of the purely theoretical celestial mechanics (see also Chap. 6). He had started to study Poincaré's recent works. In particular he had come to be very interested in the periodic solutions of the more-body problems according to Poincaré.

Jules Henri Poincaré (1854–1912)
Poincaré studied mining, mathematics and physics at the University of Paris, where he taught mathematical physics and celestial mechanics from 1881. Besides fundamental work on differential equations, philosophy and preliminary work in the field of special relativity he worked within celestial mechanics on the stability of the planetary system. A specific subject was the assessment of chaotic movements in a deterministic mechanical system. A tiny, sometimes not observable, deviation from an initial state could within a short time lead to very large changes in the development of the system. Already in the three-body problem these effects arise. He wrote his *Les Méthodes nouvelles de la mécanique céleste* between 1892 and 1899, a fundamental step forward in the theory. For his contributions to the theory of celestial mechanics Poincaré received the Bruce Medal of the Astronomical Society of the Pacific, one of the highest astronomical honours.

The first theory of Jupiter's satellites was of Laplace, after which Souillart had written an improved version, on which De Sitter's dissertation was based. But that theory was not satisfactory. There were too big differences with the observations. Some series converged too slowly, so that higher order terms, that were omitted, still had appreciable contributions. De Sitter performed new calculations and had already some preliminary and hopeful results, he wrote to Gill. Perhaps he could give a completely independent calculation of the long series expansions, because exactly in those Laplace and Souillart deviated the most [RGO, 15/130, 27-3, 29-6-1902].

De Sitter had studied the theoretical work of Hill on the system sun–earth–moon. He had not started with an elliptical orbit of the moon round the earth, a two-body problem without the influence of the sun, on which the perturbations of the sun could be calculated. He had simplified the data so that he could solve the three-body problem exactly. The other data to reach the real moon orbit, were then calculated as perturbations. In this manner he achieved much more accurate results for the orbit of the moon. This accuracy resulted also from the fact that the solution of the three-body problem was a periodical solution, giving once every certain time interval the same mutual positions.

On the basis of this example De Sitter had first resolved to start finding an exact solution of a simplified form of the four-body problem: Jupiter and its three inner moons. Later, he called the system of these orbits following from this the *orbite intermédiaire*, the intermediate orbit. The differences between the data from the intermediate orbit and the real values, the perturbations, had to be calculated from series expansions. De Sitter hoped to achieve these real values with greater accuracy and less terms. He set himself a formidable task.

Three- and four-body problem

Gill, who would become president of the British Association, invited De Sitter to be present at the annual assembly in Leicester in the first week of August 1907 [RGO, 15/130, 2-7-1906]. In April De Sitter wrote that his wife and he would go to Leicester together. Perhaps they would visit London previously for a couple of days. In all his free moments (he still worked for Kapteyn) he was busy calculating on the periodical orbits. If he had more time, he could get beautiful results. But he could, if Gill would appreciate it, tell something already about the subject at the meeting [RGS, DOG/159, 15-4-1907]. He wished to make public his first still modest results, his first steps on the road to the new theory of Jupiter's moons. In The Netherlands De Sitter's first article on the four-body solution was presented to the Academy of Sciences by Kapteyn and Ernst van de Sande Bakhuyzen, and it appeared in the *Academy Reports*. It was a contribution concerning special solutions of the three-body problem. De Sitter (1907) treated orbits of the Hestia type. Hestia is a small planet discovered in 1857. In fact De Sitter treated the system sun–earth–Hestia. Two of the three bodies he gave the masses 1 and μ, with μ much smaller than 1. The third mass (of the small planet), was in its turn much smaller than μ. He arrived at periodical solutions, families of orbits as parts of them, and the stability of the solutions according to Poincaré's methods. After this exercise he had mastered the calculatory skills, after which he could venture himself to the four-body system. As could be expected, that turned out to be not so simple and it took De Sitter until 1909 before the next step could be taken. As usual, he kept Gill informed [RGS, DOG/159, 23-5-1909]. In February 1909 the brothers Van de Sande Bakhuyzen presented to the Academy De Sitter's contribution about periodical solutions of a special case of the four-body problem (De Sitter 1909). First he considered the problem of a planet with three satellites of negligible masses, all moving in one plane, with orbital times proportional to 4:2:1. For Jupiter this is the case, if we neglect the orbital inclinations of the satellites, the influence of the fourth satellite and of the sun, and the flattening of the planet. Following Poincaré's approach of the three-body problem, De Sitter arrived at two kinds of solutions: circular orbits (first kind) and Keplerian ellipses (second kind). Next he proved that the solutions of the first kind stayed periodical even with small (non-negligible) masses, but then the condition had to be fulfilled of conjunction or opposition of the three satellites together (periodically collinear with Jupiter).

The specific situation of Jupiter and its moons makes that the movements can also be considered as a periodical solution of the second kind. Even in this case the solution can stay periodical with small but non-negligible moon masses. In the end there were 16 possible solutions, the stability of which De Sitter analysed. He proved that only one of these 16 periodical solutions is stable. And this case is realized in nature in the system of Jupiter and its three inner satellites. With this result the theoretical basis was laid for the new analytical theory.

In the meantime De Sitter had become professor of theoretical astronomy, glad after all not being director of the Observatory as well, because in that case he would never have been able to find the time to carry through these elaborate and difficult calculations. But the rest of the work, the application of his general periodical solution of the four-body system to the system of Jupiter and its three inner moons, aiming at finding for all orbital elements and masses more accurate values, again took an appreciable time. For this practical elaboration he started with the obtained general solution and he deduced from it a new set of masses in 1910, which he considered better than his own previously found masses and those of Souillart. But he could not always stay working on it and he wrote to Gill in 1912, who was delighted with the new theory, that it would still take him a lot of time to complete the theory [RGS, DOG/159, 3-1, 5-1-1912]. The complete theory would only appear in 1918, 1919 and 1925 and the processing of all available observations would take until 1929.

The New Theory

Necessary observations and World War I

Gill, after consultation with De Sitter, had written in his History of the Cape Observatory already in 1913 a summary of necessary observations for De Sitter's

Fig. 8.6 Frank W. Dyson, Astronomer Royal. Photograph from the Leiden observatory archive

programme (Gill 1913). Gill had used his authority to obtain commitments from Greenwich and The Cape for a lot of these observations, also from Dyson, then Astronomer Royal and director of the Royal Greenwich Observatory.

In January 1916 De Sitter wrote to Dyson to remind him of an earlier made promise to make a number of photographic plates [RGO, 8/88, 29-1-1916]. These, made in Greenwich, would be measured in Leiden. De Sitter wished to base his calculations on work done at the two most important English observatories: the Cape and Greenwich. Director Hough of the Cape had promised to make photographs too. De Sitter hoped that his complete programme would be ready in 1925, although that was uncertain then, in view of the outbreak of the First World War in 1914. Dyson did not respond promptly and De Sitter wrote to him again in February [RGO, 8/88, 21-2-1916]. He wanted certainty. If Dyson could not comply with his requests, he had to look further. But he preferred his English scientific friends.

In his letter De Sitter wrote elaborately about the war and his opinion about scientific freedom. Two elements from the letter are:

> I think that it is the obligation of scientists not to abandon the international contacts in this miserable war, and in particular the astronomy should set an example of continued international scientific cooperation.

I have always—even before this war—seen the danger of the German wish to dominate, especially with the smaller countries, over intellectual freedom and individuality, which is so important for the highest values in the world. In The Netherlands statesmen, scientists and businessmen have founded the patriotic club: in the interest of intellectual and economic freedom and individuality of our country; and to be alert to the great danger of Germany. The club is not in favour of the Entente,[5] but against Germany. Nobody can say how long the war will last. But from the start I have had no doubts about the outcome. And each day it becomes clearer that Germany will not be able to continue much longer. Apologies for the long letter. But the case is important.

After this letter Dyson replied quickly and promised immediately to make the photographic plates. He considered the idea of reading De Sitter's independent opinion to the meeting of the Royal Astronomical Society [RGO, 8/88, 5-3-1916]. It would take a long time before they could befriend the German astronomers again, but he would do all he could. De Sitter gave indications to Dyson how to make the plates and he remarked that the Germans had come to their senses. Fortunately not a single German astronomer had signed the declaration *Es ist nicht wahr* (It is not true) [RGO, 8/88, 13-3-1916]. In September 1914 93 German scientists, artists and authors had appealed to the cultural world (*Aufruf an die Kulturwelt*) in a pamphlet, in which all accusations against Germany were denied with a repeated *Es ist nicht wahr*. The sending of the plates was not without risks. The ship could be scuttled. Therefore they were sent via a special embassy courier, travelling from London to The Netherlands. De Sitter was pleased with Dyson's observations, but due to a new attack of his illness he could not work on them. Again, he had to take a few months rest and wrote:

God alone knows how long it will last.

[RGO, 8/88, 29-6-1916]

Endless calculations

To apply the new theory to the Jovian system, countless and complex calculations had to be performed. During the years 1916–1917 De Sitter was busy with the general theory of relativity, but after that he again worked intensively on the Jupiter system. One of his notebooks shows a lot of calculations in the period 1917–1918. De Sitter's remarks in the margin of these calculations are revealing for his efforts, doubts and successes [ASL, *Studies S10*]:

** Thus the theory has to be adapted.
** *Q.E.D.*,[6] although this is not yet correct.
** Now the proof is strict at last.
** However, a lot of difficulties remain.
** The proof as given above says nothing.
** It seems that a lot of the difficulties disappear.
** An orbit which is not an intermediate orbit but an average orbit.

---

[5]The Entente consisted of the countries opposing Germany in the First World War.
[6]Latin *Quod erat demonstrandum*: Which had to be proven.

** Then it does not lead to a solution!
** 12 January 1918: *With this the foundation of the theory is laid.*
** This is of course nonsense.
** Everything from page 97 up to here needs checking!
** The elaboration yields very long formulae.

In the meeting of the Department of Mathematics and Physics of the Academy of 23 March 1918 three remarkable subjects came up for discussion. Chairman Lorentz commemorated the deceased director of the Observatory Van de Sande Bakhuyzen. De Sitter had already been involved in the succession procedures for three weeks. Further Lorentz congratulated the member of the Academy Minister Lely[7] with the fact that his plan to dam the Zuider Zee, had been carried by parliament. At that moment Lorentz could not imagine then that he as chairman of the Zuider Zee Committee had to perform long calculations for years in order to ascertain that the provinces of North-Holland and Friesland would not be plagued by high water levels after the construction of the dam. The third point was the presentation by De Sitter of his 'Outlines of a new theory of Jupiter's satellites', of which the foundation had been completed in January. A next part would follow soon. These outlines would also be published in the *Annals of the Observatory in Leiden* (De Sitter 1918), as first part of a series of three. The second part of the theory, in the form of a dissertation by his student A. J. Leckie, would follow in 1919. The third part, again by De Sitter would follow in 1925.

First part: Outlines of a New Mathematical Theory of Jupiter's Satellites
De Sitter's results in solving the four-body problem made it meaningful to take the Jupiter system with the three inner moons as a starting point. The influence on the periodical solutions of this system by the fourth satellite could be dealt with as a perturbation, together with the other perturbations, those by the sun and by the flattening of the planet. However, De Sitter made an intermediate step. He constructed a new basis including the fourth moon. To that end he had to calculate a number of variations and add them to the first solution. After that the remaining influences had to be treated as perturbations. An important advantage of this intermediate orbit with variations was the absence of small divisors, which gave great problems in the regular theory.[8] As another relevant difference the plane of the equator of Jupiter was taken as plane of reference, instead of the plane of the orbit of Jupiter round the sun.

Second part: Analytical and Numerical Theory of the Motions of the Orbital Planes of Jupiter's satellites

---

[7]Cornelis Lely (1854–1929) studied for civil engineer in Delft. He became Minister of Water Management in three different governments, the third time from 1913 until 1918, when he steered his plan to dam the Zuider Zee through parliament. The IJsselmeer Dam was closed in 1932. The Zuider Zee, part of the North Sea, became a lake: the IJsselmeer, with an area of about 2000 km^2, of which half has been reclaimed since.

[8]Small divisors give large values of a division, which required then many more terms in the series expansions.

The second part consisted of the dissertation of A.J. Leckie (1919). The mutual proportions as whole numbers of the orbital times of the three inner moons did not play a relevant role in the determination of the inclinations and the nodes. To start with, those could then be evaluated. At the end of his dissertation Leckie compared his results with those from Souillart's theory and gave an analysis of the differences.

August Juliaan Leckie

Leckie was born in Paramaribo, then a colony of The Netherlands. After the death of his father his mother had to care for eight children alone. Leckie received a scholarship for a couple of years. He would do his bachelor's exam in 1911 [AUL, Faculty minutes]. Already in 1916 he worked on the inclinations of Jupiter's satellites according to De Sitter's new theory [ASL, *Studies S10*]. He received his doctor's degree in 1919.

De Sitter published the definitive orbital elements and masses of the moons in 1929. In the preface of his dissertation Leckie thanked the deceased Van de Sande Bakhuyzen and in particular De Sitter, for invaluable help during his research, and for the interest often shown outside his studies. So the interest went further than only the scientific results. Later, around 1930, Leckie worked in the Dutch East Indies as first assistant of H. R. Woltjer, who had succeeded Jacob Clay[9] as director of the Bandung Institute of Technology on the island of Java. This Institute had, like the Observatory at Lembang, near Bandung, been founded with financial support of the Dutch tea tycoon K. A. R. Bosscha (1865–1928). Woltjer considered Leckie a very capable researcher, meticulous and very energetic. Among other subjects he published on atmospheric physics (Pyenson 1989).

Third part: New mathematical Theory of Jupiter's Satellites

De Sitter's publication of his new mathematical theory (De Sitter 1925) determined the intermediate orbit and the variations. In March and June 1919 he had already presented two articles to the Academy: one about the intermediate orbit and one about the variations, but only the variations of the first degree. In the meantime De Sitter had completed the calculations with the second and third degree variations, and the series expansion in powers of the corrections on the assumed masses. But all was not yet ready. Work on a number of perturbations that had not yet been taken into account was still in progress and would be published later.

The consequences of the proportion in whole numbers of the three orbital times, the cause of the bad convergence in the old theory, had now been taken into account in the first approximation. It is true, the intermediate orbit with variations did not

---

[9]The Dutch physicist Jacob Clay (1882–1955) studied physics with Kamerlingh Onnes and Lorentz in Leiden. He was professor in Bandung from 1920 until 1929. On the way back to The Netherlands he discovered that cosmic radiation probably consisted of charged particles.

represent a possible natural set of movements, but an approximation of the real movements. And further, De Sitter had introduced numerical values for the average movements of the satellites and the average masses from the start, so that a limit could be determined to the accuracy, and the certainty could be incorporated that no terms would be skipped which exceeded the fixed inaccuracy. De Sitter continued in the calculations of the expansion terms up to and including a full additional decimal in the relevant quantities above the best data in his time. Thus his theory could last for a long time. To be sure he had calculated all intermediate results one or two decimals further. He had performed part of these calculations during his stay in the sanatorium in Arosa in Switzerland.

This was what it had been all about for De Sitter all those years. The winding up of *his* new theory. He had worked on it for a quarter of a century, in the meantime he had defeated a wasting disease, salvaged an observatory from decline, made a career in the government of the University and in the International Astronomical Union (see further chapters). But the presentation of this manuscript must have been a highlight of his life. The calculatory work on it had been a multiple of his work on the cosmological models nearly ten years earlier (see next chapter), even in complexity. He had reached his mathematical and theoretical summit. The new theory based on his solution of the four-body problem was an advancement in the analytical celestial mechanics. From 1925 onwards his name belonged in the row Laplace, Souillart, Hill, Poincaré. There are no documents about it, but it will have given him probably more satisfaction than his solution of the Einstein equations of the general theory of relativity. De Sitter became world-famous in the footsteps of Einstein, but in the sequence of his achievements he would very probably have put the new theory of Jupiter's satellites on number one.

Old and new observations and publications
Many more publications on Jupiter's satellites would follow. A lot of observations from the distant past, as far back as the seventeenth century, and also very recent ones, often made at the request of De Sitter at several observatories like Cape Town, Greenwich and Leiden, would later be reduced by his students and himself, in order to determine with increasing accuracy numerical constants and the data of Jupiter's satellites.

Before the publication of De Sitter's final article Pieter Kremer, a student of De Sitter, had received his doctor's degree in the framework of this research programme in 1923 (Kremer 1923).

Later, De Sitter's friend Innes from the Union Observatory in Johannesburg and Harold Alden (1890–1964) of the southern station of the American Yale Observatory in South Africa, where Schlesinger was director, also cooperated [ASL, 26-4 to Alden, 9-5 from Schlesinger, 1927].

In 1929 a sizeable article appeared in the *Annals of the Observatory in Leiden*, containing a discussion of old eclipses of Jupiter's satellites (De Sitter 1929). In particular there were still uncertainties about the orbital elements of satellites III and IV. The movement of the perijovum of satellite III, the position on the orbital ellipse nearest to Jupiter, was calculated on the basis of measurements covering a period of only thirty years. That was not yet a quarter of a complete revolution of the perijovum (the elliptical orbit revolves slowly around Jupiter). For an accurate determination observations over a period of a century were necessary. De Sitter searched for old observations, but he came to the conclusion that all the effort that would be necessary for the reduction of these old observations would be out of proportion compared to the expected results. He decided to take on only those old observations, which already had been compared to the existing tables. In this manner he could restrict the amount of work drastically. Among others he took eclipse observations from the period of 1673 until 1896, that had been processed by the Swedish astronomer Pehr W. Wargentin (1717–1783) and Sampson. For that end they had used the tables of Damoiseau.

De Sitter calculated for satellite III a rotation of 0.007084° per day of the line of nodes and the perijovum, while the modern observations had yielded 0.007216° per day. The first value, based on the old observations, corresponded with a full rotation round Jupiter in 139.2 years and the second, based on the recent observations, with a full rotation in 136.7 years. Using the old eclipse observations, De Sitter had gained an appreciable adaptation.

Finishing the Jupiter work
The years 1927, 1928 and 1929 were the last in which De Sitter published on the moons of Jupiter. The last observations from Gill's programme had been made at the Cape in 1924. A publication with the results for the lengths of the moons of Jupiter appeared in the *Bulletin of the Astronomical Institutes of The Netherlands* (De Sitter 1927a), followed by an extensive treatment in the *Annals of the Observatory* (De Sitter 1928). Part of this publication (Brouwer 1928) in the *Annals* was written by Dirk Brouwer, who had received his doctor's degree with De Sitter based on this work in 1927 (Brouwer 1927). In fact, with these last articles De Sitter's aim was reached: the determination of a set of data of Jupiter's satellites, having made use of nearly all existing observations, which had been made at the most favourable times and with the optimal methods. It was De Sitter's conviction that the determined orbital elements could not soon be replaced by more accurate ones.

On 8 May 1931 De Sitter held the Darwin[10] Lecture (De Sitter 1931). Delivering this lecture belonged to the reception of the Gold Medal of the Royal Astronomical Society, the greatest honour that could be bestowed on an astronomer. In his lecture he treated the complete history of observations and theories concerning the four big satellites of Jupiter.

In the last part of the lecture De Sitter also discussed extensively the weak points and the still unsolved problems. A number of appreciable residues in the lengths of the satellites rested without a proper explanation by the theory. A tricky point was the measuring of time. Astronomers had always started from the supposition that the rotation of the earth was stable and that it was therefore a perfect measuring instrument of time. But from many measurements, by Innes and Brown among others, it had become all the time more clear that the earth changed its speed of

---

[10]The astronomer George H. Darwin (1845–1912), son of Charles Darwin, was president of the RAS from 1899–1901.

rotation intermittently. When one would not take into account these sudden changes, all kinds of non-existing effects would appear in the measurements of time dependent properties of celestial bodies. But even when one did take into account these variations in the rotation of the earth, there remained unexplained differences, that were not within the limits of the measuring errors.

The libration stayed a problem too. De Sitter referred to the amplitude of $0°.158$, found by him in 1907. The value of the amplitude based on the available data in 1928 was in the meantime six times smaller: $0°.025$. But even this value had a large margin of error. The conclusion had to be that the libration, if it existed at all, was too small to be measured at that moment. But he added something notable. His student Brouwer had found exactly the same values as De Sitter on the basis of eclipse measurements from Johannesburg. Brouwer too came to the conclusion that actually the effect of libration was too small to be measured with certainty. De Sitter said that these perfectly matching results, based on completely different methods, was quite remarkable. Honesty commanded De Sitter to ascertain the probable imperceptibility of the libration, although that must have saddened him. It is remarkable that he mentioned Sampson only once, in a table. Sampson had received the Gold Medal in 1928, also for his work on the satellites of Jupiter. However, in the lecture in which the RAS awarded Sampson the Medal it was said that it was slightly surprising he had only used eclipse measurements. In contrast to Sampson De Sitter had used measurements, obtained with a broad variety of methods. Besides he had made a formidable contribution to the theory. De Sitter had to accept the remaining unexplained residues. His work on Jupiter's satellites had come to an end. He could look back on it with pride. Two years earlier, aboard the Nestor sailing to South Africa to attend the meeting of the British Association, he had asked Dyson if he should continue working on the moons of Jupiter. Dyson's answer was: *don't do it.*[11] It had been good, and enough.

A few of De Sitter's students
Jan Woltjer Jr. (1891–1946)
Woltjer studied in Amsterdam and Leiden. From 1913 onwards he worked at the Observatory in Leiden. He received his doctor's degree with De Sitter as supervisor in 1918 on his dissertation *Investigations in the Theory of Hyperion*, Saturn's eighth moon. He became assistant, lecturer and professor of theoretical astronomy, at first to relieve De Sitter after his appointment as director of the Observtory. After 1925 he transferred his field of research from celestial mechanics to the theory of star atmospheres and the inside of stars. He died as a consequence of undernourishment during the so-called winter of starvation[12] 1944–1945 at the end of World War II (Brouwer 1947).

---

[11]ASL, Conversation between De Sitter and Dyson, sailing to South Africa, 6 July, 1929.
[12]Dutch: Hongerwinter.

Dirk Brouwer (1902–1966)

Still being a student in Leiden Brouwer determined the mass of Saturn's moon Titan from its perturbations on the other moons. He was coached by Woltjer during his studies. In 1927 he received his doctor's degree, supervised by De Sitter (Brouwer 1927). In the preface Brouwer thanked De Sitter for the responsibility given to him to compare Innes' excellent measurements to a refined theory. It had given him great satisfaction. Subsequently Brouwer worked in Berkeley (California, USA) for a year, after which he was appointed at Yale University (New Haven, Connecticut, USA). He was appointed there as a professor in 1941 and stayed there until his death in 1966. Together with G. M. Clemence (1908–1974) Brouwer wrote a textbook on celestial mechanics, from which a whole generation of students learnt the subject. He was one of the first astronomers to use the computer for his extensive calculations. He calculated the orbits of the first artificial satellites, and the other way round he could gain more knowledge on the form and composition of the earth from the data of the satellites. In a review article on celestial mechanics from 1963 (Brouwer 1963) Brouwer still referred in the chapter on the theory of artificial satellites to a method of his old master De Sitter to determine the form of the earth (De Sitter 1924). Brouwer received the Gold Medal of the RAS in 1955 and posthumously the Bruce Medal of the Astronomical Society of the Pacific in 1966. Two prizes bear Brouwer's name: one from the American Astronomical Society for achievements in a broad field, including dynamical astronomy and celestial mechanics, and a second one from the American Astronautical Society, among other subjects for important contributions to the mechanics of space travel.

Simon Cornelis van Veen (1896–?)

Van Veen worked from 1919 until 1921 as an assistant under Ehrenfest. He wrote his dissertation with De Sitter in 1927. It was a theoretical dissertation and dealt with commensurability[13] of the mean movements of planetoids (or asteroids) with that of Jupiter: *Periodic Solutions in Commensurability Areas, with application to the problem of the lacunae in the system of the asteroids.*[14] Kirkwood[15] had stressed the point that there were hardly planetoids at the proportions 1/3, 3/7, 1/2 and 2/5. Except for a few exceptional cases there was not yet a complete theory. This work was closely related to De Sitter's own theory concerning the intermediate orbit. In his introduction Van Veen

---

[13]Commensurability means proportions in whole numbers. For example: the three inner moons of Jupiter have for their the orbital times the commensurability of 1:2:4.

[14]Dutch: *Periodieke Oplossingen in Commensurabiliteitsgebieden, met toepassing op het probleem der lacunes in het stelsel der asteroïden.*

[15]Daniel Kirkwood (1814–1895) was an American astronomer, who worked mainly on asteroids. He discovered the later called Kirkwood gaps in their distances to the sun, in his opinion related to orbital resonances with Jupiter.

thanked in particular Ehrenfest and Lorentz. De Sitter's role seemed to have been restricted to the official status of supervisor, while Woltjer was particularly thanked for his permanent coaching. Van Veen was teacher at the hbs in Dordrecht (province of South-Holland) from 1921 until 1946. In that year he was appointed professor at the Technical University in Delft.

De Sitter and Brouwer on Astronomical Constants

At an international congress in Paris in 1896 a large number of astronomical constants was fixed to be used in astronomical calculations. They were to a large extent based on work by Newcomb. Before that time appreciably differing values were used, which often were not even logically connected. But new observations led to better values and there were also deficiencies in Newcomb's set of values. De Sitter had encountered these problems in his work too. He published a modest article on a number of constants in 1927 (De Sitter 1927b). De Sitter did not aim at completeness, and he considered the time not yet appropriate for a revision of the in the meantime thirty years old standard work of Newcomb. A few of the values calculated from De Sitter's most recent observations were:

** Mean radius of the earth: 6371.238 ± 0.030 km;
** Gravitational acceleration at the mean latitude: 979.770 ± 0.002 cm/s^2;
** The constant of precession[16]: 50.2486 ± 0.0010″/y;
** Sun parallax (radius of the earth/distance earth–sun) 8.8032 ± 0.0013″.

De Sitter continued studying the astronomical constants in the ensuing years, but he died before he could publish again on the subject. At the request of the later director of Leiden Observatory Jan H. Oort Brouwer wrote a sequel to the first article, based on De Sitter's data. De Sitter was a posthumous author (De Sitter 1938). The aim of De Sitter had been to determine a system of values, in which every theoretical relation between the quantities was satisfied, and where for each constant a value was adopted lying within the margins of error of the observations. Brouwer performed this task in a commendable manner, indicating precisely where he had put in corrections and changes in De Sitter's original manuscript and preliminary notes. Oort and Brouwer knew how important De Sitter had considered working with the best astronomical constants. From these investigations it appeared

---

[16]The rotation axis of the earth describes a complete turn in nearly 26,000 years (=360.60.60″/ 50.2486″).

again that beside theoretician De Sitter was an excellent practical astronomer too. Brouwer bestowed on his master the appropriate honour with this article. Later, in 1965, he published again on the relation between several constants, having a different, sometimes simpler form (Brouwer 1965).

Later Observations

It is remarkable how good the calculations of the masses of Jupiter and its satellites have been made by Sampson and De Sitter, in view of later observations. New results were found with observations made by the two Pioneer spacecraft, which were compared to their old results. Pioneer 10 was launched in 1972, Pioneer 11 in 1973, reaching Jupiter in 1973 and 1974 respectively, after crossing the planetoid belt. They did extensive observations on Jupiter and its satellites. The influence of the moons and of Jupiter itself on the Pioneers' orbits gave accurate information on the masses (Burns 1977).

The tables for the satellites of Jupiter, made by Sampson in 1910, showed after more than half a century differences making a new theory necessary. In the 1970s of the twentieth century work was done on the revision of Sampson's work and suggestions were made to develop further De Sitter's elegant theory.

The problem of the libration had not yet been solved around 1980. According to J. H. Lieske in *Satellites of Jupiter* (Lieske 1982) the amplitude of the libration was $0.066°$ with a period of around 6 years. Other scientists in those years supposed an even lower value of the amplitude. Although, this value was more than twice the uncertain value of De Sitter and Brouwer.

In 1977 NASA launched two Voyager spacecraft, 1 and 2, which reached Jupiter in 1979 and went on to Saturn, Uranus and Neptune. They took more than 50,000 photographs of Jupiter and its moons. For the first time observations could be made of the astrophysical aspects of the planet and its satellites (e.g. active volcanoes on Io).

In 1989 the appropriately named spacecraft Galileo was launched to Jupiter, arriving there in 1995, going into orbit round the planet. Its observations supported the theory of liquid oceans under the surfaces of Europa, Ganymede and Callisto. At the end of its mission Galileo was sent into Jupiter's atmosphere in 2003.

In 2011 spacecraft Juno (in Roman mythology the wife of Jupiter, mistress and master of the universe) was launched, arriving at Jupiter in 2016, where it went into orbit round the planet. It will perform an extensive programme of measurements of a vast number of chemical, meteorological, geological and physical properties of Jupiter. It will be sent into Jupiter's atmosphere probably in 2018, in order to eliminate space debris.

The big differences between the material composition of the satellites provides information about the history and development of the Jovian system. Thanks to this space exploration a completely new and successful research programme of the planet and its moons has been going on since the last quarter of the twentieth century. Perhaps De Sitter himself has sometimes dreamt of so much detailed information about *his* satellites, about the *guards* of Jupiter (Dutch: *wachters*), in the words of his father.

References

Bottlinger, C.F. (1912). 'The explanation of the empirical parts of the movement of the moon by the assumption of an absorption of the gravitation in the inner parts of the earth' (German: 'Die Erklärung der empirischen Glieder der Mondbewegung durch die Annahme einer Extinktion der Gravitation im Erdinnern'); *AN*, 191, p. 147, 1912.

Brouwer, D. (1927). *Discussion of the observations of satellites I, II and III of Jupiter, made at Johannesburg by dr. R.T.A. Innes in the years1908–1925* (Dutch: *Diskussie van de waarnemingen van satellieten I, II en III van Jupiter, gedaan te Johannesburg door dr. R.T. A. Innes in de jaren 1908–1925*), E. IJdo, Leiden, 1927.

Brouwer, D. (1928). 'Discussions of observations of Jupiter's satellites made at Johannesburg in the years 1908–1926'; *AOL*, XVI, Part 1, 1928.

Brouwer, D. (1947). 'Obituary J. Woltjer Jr.'; *MNRAS*, 107, pp. 59–60, 1947.

Brouwer, D. (1963). 'Review of Celestial Mechanics'; *Annual Review of Astronomy and Astrophysics*, 1, pp. from 219, 1963.

Brouwer, D. (1965). 'Relations among some important astronomical constants, The system of Astronomical Constants'; *Proceedings of the IAU, edited by Jean Kovalevsky*, pp. from 241, Gauthier-Villars, Paris, 1965.

Burns, J.A., editor, (1977). *Planetary Satellites*, The University of Arizona Press, Tucson (Arizona, USA), 1977.

De Sitter, W. (1907). 'About periodical orbits of the Hestia type' (Dutch: 'Over periodieke banen van den Hestia-typus'); *NAW2*, XVI, pp. 35–44, May 1907.

De Sitter, W. (1909). 'About the periodical solutions of a special case of the four-body problem' (Dutch: 'Over de periodieke oplossingen van een speciaal geval van het vier-lichamenvraagstuk'; *NAW2*, XVII, pp. 752–769, February 1909.

De Sitter, W. (1911). 'On the Bearing of the Principle of Relativity on Gravitational Astronomy'; *MNRAS*, LXXI, 5, pp. 388–415, 1911.

De Sitter, W. (1912). 'Absorption of gravitation'; *The Observatory*, XXXV, 454, pp. 387–393, 1912.

De Sitter, W. (1913a). 'Absorption of gravitation and the length of the moon' (Dutch: 'Absorptie van gravitatie en de lengte van de maan'); *NAW2*, 21, pp. 737–752, pp. 1019–1035, 1913.

De Sitter, W. (1913b). 'Ein astronomischer Beweis für die Konstanz der Lichtgeschwindigkeit'; *PZ*, 14, p. 429, 1913.

De Sitter, W. (1918). 'Outlines of a New Mathematical Theory of Jupiter's Satellites'; *AOL*, XII, Part 1, 1918.

De Sitter, W. (1924). *BAN*, II, pp. from 109, 1924.

De Sitter, W. (1925). 'New mathematical Theory of Jupiter's Satellites', *AOL*, XII, Part 3, 1925.

De Sitter, W. (1927a). 'Summary of the results derived from a discussion of the longitudes of Jupiter's satellites, with special reference to the rotation of the earth'; *BAN*, III, pp. 247–257, 1927.

De Sitter, W. (1927b). 'On the most probable values of some astronomical constants, first paper, constants connected with the earth'; *BAN*, IV, pp. 57–61, 1927.

De Sitter, W. (1928). 'Orbital elements determining the longitudes of Jupiter's satellites, derived from observations'; *AOL*, XVI, Part 2, 1928.

De Sitter, W. (1929). 'Discussion of Old Eclipses of Jupiter's Satellites'; *AOL*, XVI, Part 4, 1929.

De Sitter, W. (1931). 'George Darwin Lecture, delivered by Professor W. de Sitter, Assoc. RAS, on 1931 May 8'; *MNRAS*, XCI, 7, pp. 706–738, 1931.

De Sitter, W. (1933). 'In Memoriam'; *Leidsch Dagblad*, 26 September, 1933.

De Sitter, W. (1938). 'On the system of astronomical constants (edited and completed by Dirk Brouwer, with the aid of notes by W. de Sitter)'; *BAN*, VIII, pp. 213–231, 1938.

Gill, D. (1913). *A History of the Royal Observatory Cape of Good Hope*, His Majesty's Stationary Office, London, 1913.

Kremer, P. (1923). *Discussion of Micrometer Observations of Jupiter's Satellites, Done in Washington in the years 1903–1906* (Dutch: *Discussie van Micrometerwaarnemingen van de Satellieten van Jupiter, Gedaan in Washington in de jaren 1903–1906*), Eduard IJdo, Leiden, 1923.

Leckie, A.J. (1919). 'Analytical and Numerical Theory or the Motions of The Orbital Planes of Jupiter's Satellites: Secular Terms'; *AOL*, XII, Part 2, 1919.

Lieske, J.H. (1982). In: *Planetary Satellites (D. Morrison, editor)*, The University of Arizona Press, Tucson (Arizona, USA), 1982.

Pyenson, L. (1989). *Empire of reason*. See References of chapter 1.

Röhle, S. (2007). *Willem de Sitter in Leiden - A chapter in the History of the Reception of the Relativity Theories*, dissertation (German: *Willem de Sitter in Leiden - Ein Kapitel in der Rezeptionsgeschichte der Relativitätstheorien*), Johannes Gutenberg-Universität, Mainz, 2007. The dissertation of Röhle is a gold mine of relevant information about De Sitter.

Chapter 9
Gravity, General Relativity and Cosmology

The supposition that the physical space is in its smallest parts identical with the Euclidian one, is a hypothesis which is absolutely not based on experience (Proposition XII in De Sitter's first dissertation on Mathematics and Astronomy).

Although the law of inertia has only been found 'ganz nebenher' (casually, according to Ernst Mach*) by Galileo, it must still be considered as the most important result of his discoveries* (Proposition XVII in De Sitter's second dissertation on Mathematics and Physics).

"Space seems to be highly curved in this region."
Arthur Eddington after a failed golf hit

(Vibert Douglas 1956)

A Few Elements from Einstein's General Theory of Relativity

In all his theoretical investigations De Sitter was always interested in the fundamental questions. In the short propositions mentioned above from his dissertations in 1901 we already read about his interest for the geometrical properties of space and the fundamental concept of inertia. Gravity had fascinated De Sitter from the start of his career. In his inaugural lecture in 1908 he said that all attempts to penetrate into the internal mechanism of gravity had failed. It was Albert Einstein (1879–1855) who would give the solution with his general theory of relativity. This paragraph describes some of the aspects of this theory in a nutshell (Janssen and Lehner 2014; Pais 1982, 2005).

© Springer Nature Switzerland AG 2018
J. Guichelaar, *Willem de Sitter*, Springer Biographies,
https://doi.org/10.1007/978-3-319-98337-0_9

Relativity Principle

Einstein's Special Theory of Relativity only considered co-ordinate systems moving with a constant velocity relative to each other. In those co-ordinate systems all laws of nature had to have the same form, had to be covariant. According to Einstein this also had to be the case in co-ordinate systems moving with an arbitrary acceleration relative to each other. This was the Relativity Principle.

The Coil and the Magnet

Thinking about the movements of a coil and a magnet in an electromagnetic field brought Einstein at last in 1907 to what he called the most happy thought of his life. If you move a magnet to a circular copper winding at rest, an electric current in the winding is generated, an induction current. But if you, the other way round, move the copper winding to a magnet at rest, the electrons in the winding experience a Lorentz force and they start moving around in the winding: also an electric current. But the currents occur in different co-ordinate systems. In Einstein's view these could not be two fundamentally different effects. In the end it had to be possible to describe them both in the combined electromagnetic field. The realization that it was one and the same effect in two different co-ordinate systems brought Einstein to what he called the happiest thought of his life.[1]

The Equivalence Principle

Just like the fact that the coil and magnet described only one physical process, in different co-ordinate systems, the idea came to Einstein to consider a woman in a cabin without a window.

Case 1. Let the cabin with the woman fall free under the gravitational pull of the earth. In the co-ordinate system of the earth the woman and the cabin fall with exactly the same acceleration to the earth, meaning that the woman is floating weightlessly inside the cabin. But let the cabin and the woman be far from the earth in space and far from the gravitational influences of other masses. Then the woman will also float freely within the cabin.

---

[1]Einstein wrote down these thoughts in an unpublished article in 1920 (*Fundamental Ideas and Methods of the Theory of Relativity, Presented in Their Development*), intended for *Nature*. The elements of Einstein's search to find the general theory of relativity are described in a clear manner by Michel Janssen in *'No Success Like Failure ...': Einstein's Quest for General Relativity, 1907–1920*; Colloquium, Astrophysics Group, Michigan State University, April 7, 2011.

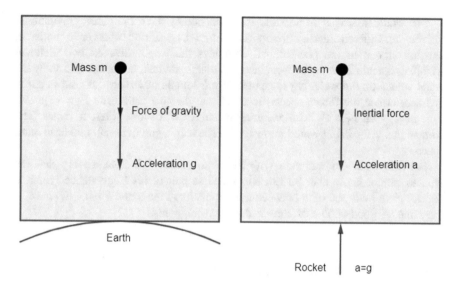

Fig. 9.1 The Equivalence Principle

So she does not know the difference between being in free fall in the gravitational field of the earth or being somewhere far in outer space.

Case 2. Let the cabin with the woman be at rest in the gravitational field of the earth. The woman is pressed on the floor of the cabin by the gravitational force of the earth. But let the cabin be in outer space with a completely noiseless rocket engine giving the cabin a constant acceleration. The floor of the cabin presses the woman to give her the same acceleration.

Again she does not know the difference between being at rest in a gravitational field or accelerated in outer space.

This *Gedankenexperiment* or thought experiment led Einstein to formulate his Equivalence Principle:

> The gravitational field has a relative existence only, in a manner similar to the electric field generated by electro-magnetic induction, dependent of the acceleration of the co-ordinate system.

Gravitational Mass and Inertial Mass

Gravitational mass is the measure of the gravitational force F_g, which a gravitational field exerts on a body, giving it an acceleration g: $F_g = m_g g$. The constant of proportionality is the gravitational mass m_g.

Inertial mass is the resistance of a body, with which it opposes a change of velocity. If a force F works on a body, giving it an acceleration a, the second law of Newton holds: $F = m_t a$, in which we call m_t the inertial mass.

You could ask if it is possible that two bodies have the same gravitational masses, but different inertial masses (take $m_{t1} > m_{t2}$). If our observer in the cabin hanging at rest in the gravitational field lays the two bodies on two identical weighing machines on the bottom, they feel equal gravitational forces and show the equal values on the weighing machines. If our female observer is far out in space and accelerated the forces needed to accelerate the two bodies are $F_1 = m_{t1}a$ and $F_2 = m_{t2}a$, with $F_1 > F_2$. The weighing machines will show different values. The woman inside the cabin would know then if she is in a gravitational field or in outer space.

But this is in contradiction with the Equivalence Principle, stating that the woman cannot know that. So Einstein stated as part of his Equivalence Principle that for each body the ratio between the gravitational mass and the inertial mass is the same. A good choice of units makes then: $m_t = m_g$.

Ehrenfest Paradox and Curved Space

For Einstein it was a many years' struggle to find the right road to general relativity. Ehrenfest played a role in this process. From St. Petersburg he published an article of only half a page in the *Physikalische Zeischrift* in 1909 about uniform rotation of rigid bodies and relativity theory (Ehrenfest 1909). In this article he described a paradox, later called the Ehrenfest Paradox. Ehrenfest considered a rotating rigid cylinder. For the observer on the rotating cylinder R' is the radius and for the perimeter P' we have $P' = 2\pi R'$. For an observer at rest outside the rotating cylinder we have in the first place that the radius R, of which each small part moves perpendicular to the velocity, has no length contraction, so $R = R'$. If the usual formula for calculating the perimeter of a circle would hold, the observer outside the cylinder would have: $P = 2\pi R = 2\pi R' = P'$. But now comes special relativity into action. Each small part of the perimeter moves in the direction of the velocity. The observer outside then measures a length contraction, from which follows $P < P'$. There we have the paradox. There seemed to be something wrong with the formula $P = 2\pi R$ in the flat plane. In the literature this paradox has been discussed many times. Einstein and Ehrenfest discussed the paradox during Ehrenfest's visit to the Einstein family in Prague in the beginning of 1912. It was the start of their friendship, with long talks about physics, and making music on the violin (Einstein) and at the piano (Ehrenfest). The talks in 1912 about the paradox and other subjects supported Einstein to set further steps in his quest for a theory of gravitation. Something seemed not to be in order with the space. During the talks the question must have arisen: is our flat 3-dimensional space curved within a 4-dimensional space?

In the next example we encounter a similar case, in which the 2-dimensional space is curved in a 3-dimensional space. Imagining a 2-dimensional world on a global sphere. There are no problems on a very small scale on the sphere with measuring the perimeter of a circle: $P' = 2\pi R'$. But on a large scale there are. Let us

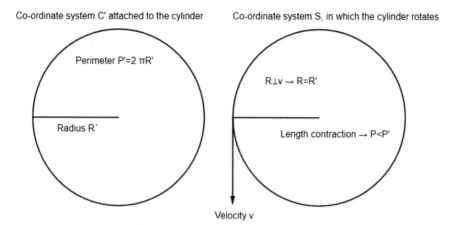

Fig. 9.2 The Ehrenfest paradox

take the extreme circle on the sphere: the equator. The radius R on the sphere we have to take is along a meridian line from the North Pole (the centre of the circle) to the equator. For the Perimeter P we have the formula P = 4R, with 4 instead of 2π. The normal formula would yield P' = 2πR', with P' > P. The difference comes of course by the curved 2-dimensional surface of the sphere in the 3-dimensional space.

So thinking of a theory using curved space was a definite option.

Bending of Light Rays

The bending of a light ray passing the sun at a small distance may be understood qualitatively making use of the Equivalence Principle only. Suppose we have a cabin with an observer with mass m, who cannot look outside, standing still relative to the sun. The observer feels the gravitational force mg, with g the gravitational acceleration of the sun. Suppose a second situation, in which a cabin far from the earth is accelerated with a constant acceleration a by a noiseless rocket, with a = g. The observer inside the cabin far away feels the same force as the one at rest near the sun. The Equivalence Principle says that the observer cannot decide which is the case for her. Suppose now that a light ray enters the cabin horizontally (parallel to the floor) through a tiny hole. In the case of the accelerated cabin this light ray will reach the other side of the cabin slightly nearer to the floor. It seems to have made a bend. The Equivalence Principle states now that the observer in the cabin at rest relative to the sun cannot observe a difference. The conclusion is that in the cabin near the sun the light ray makes the same bend, now as a result of being in a gravitational field.

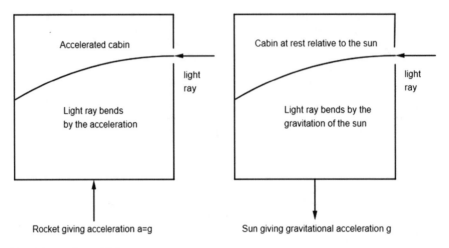

Fig. 9.3 Bending of light rays

This bending was not easy to measure. But shortly after the publication of Einstein's theory in 2015 starlight passing near the sun during a total eclipse made it possible to measure the bending during two expeditions in 1919. The bending was in agreement with the prediction of Einstein's theory.

The Einstein Equations

Making use of curved space was at last the road for Einstein to arrive at his new theory. He made use of the complex non-Euclidian geometry, developed by the German mathematician Bernhard Riemann (1826–1866) in the nineteenth century. A friend of Einstein, the mathematician Marcel Grossmann (1878–1936), helped him in applying this difficult mathematics to solve his gravitational problems.

> Grossmann, you must help me, otherwise I will become insane.

Einstein told him (Pais 1982, 2005). An event in spacetime has four co-ordinates x_μ (μ = 0, 1, 2, 3, with 1, 2, 3 for the space co-ordinates and 0 for the time). The curvature of the 4-dimensional spacetime was described by Einstein and Grossmann with the help of the so-called metric tensor $g_{\mu\nu}$ (μ, ν = 0, 1, 2, 3) consisting of 16 functions of time and place x_μ. The metric tensor describes the curved spacetime. The mathematics was already difficult enough. But by what was the metric determined? It is clear that what is inside space, the masses, their energies, velocities and impulses, influence the space they take up and also the space around them. In simple words it can be stated as follows: the energy-impulse determines the curvature (metric) of spacetime and the metric in its turn determines the movements of the masses and light rays. Of course these things happen according to laws of nature. It were these laws and equations Einstein was looking for.

In the beginning of 1914 Einstein and Grossmann published an *Entwurftheorie* (draft theory), although it was not yet a complete theory. But, after having overcome still many more problems, Einstein managed to write down the definitive field equations and publish them at the end of 1915. The left-hand side of the equation contained $G_{\mu\nu}$, containing the metric $g_{\mu\nu}$ in a very complex form, which is well beyond the realm of this book. The right-hand side contained the several components of the masses, the energies and the impulses, together in the energy-impulse tensor $T_{\mu\nu}$. Very schematically the Einstein equations have the form:

$$G_{\mu\nu} = -\,\kappa\,(T_{\mu\nu} - \tfrac{1}{2}Tg_{\mu\nu})$$

The constant κ relates the left-hand side to the right-hand side of the equation: spacetime to mass and energy (T is a special combination of the values of $T_{\mu\nu}$). For the proportionality constant we have:

$$\kappa = 8\pi G/c^4,$$

in which G is the constant of gravitation and c the speed of light. In the approximation to a metric describing a flat space, the normal Newtonian law of gravity could be deduced.

Finding solutions of the Einstein equations was and is difficult and very complicated. Einstein himself calculated at the end of November 1915 using his new equations the missing rotation of the perihelion of the planet Mercury. He found 43″ per century, which was within the measuring errors the correct answer. It was the first success of the general theory of relativity. Einstein was overjoyed for days. The Leiden group, with Lorentz, Ehrenfest and De Sitter, had followed closely the development of the new theory. De Sitter would master the general theory of relativity as fast as he had the special theory.

The General Theory of Relativity Conquers Leiden

The theoretical physicists in Leiden, in particular Lorentz and Ehrenfest, and astronomer De Sitter followed the developments of the general theory of relativity accurately, also in the period in which Einstein and Grossmann took the geometry of curved space as a vehicle for the new theory and published the *Entwurftheorie* in January 1914 (Kox 1993). After a lecture by Lorentz at Teyler's Museum in Haarlem in that month De Sitter made a few notes on Einstein's preliminary gravitation theory in one of his *Studies* notebooks. He thought some elements *absolutely incomprehensible* and asked himself:

But is this Einstein's last word?

Nevertheless, on the basis of the preliminary theory, he made some calculations on the perihelion shifts of the planets and on the movements of the moon. Like many physicists of those days De Sitter struggled with the fundamental concepts. He noted:

In any case Einstein's equations are very complicated and really physical: utterly impracticable for astronomical problems.

He was very critical of the fact that the field equations were given relative to an absolute co-ordinate system, and that they were not given from one body relative to another. He considered the inclination using a field and not a force anti-relative and borrowed from the ether theory [ASL, *Studies S2*].

At the end of November 1915 Einstein published the definitive version of the field equations of his general theory of relativity. Already at the end of December De Sitter calculated as a finger exercise the perihelion shift of Mercury, like Einstein had done shortly before, arriving at the correct value of 42″,9. Slightly further in his notes he came to the conclusion that *nothing remained* of Seeliger's theory, with its rotating dust clouds inside the orbit of Mercury [ASL, *Studies S12*]. De Sitter plunged with his whole mathematical and astronomical soul into the study of the new theory. He had mastered much earlier the special theory of relativity and knew already the older versions of the general theory of relativity. Moreover, he attended as a *student* the Monday morning lectures of Lorentz about the new theory, from March until June of 1916. De Sitter managed, supported by all his work over the past years in trying to grasp gravity, to master fast the complex calculation techniques and to develop ideas on the astronomical implications. Then those ideas were on his mind constantly. It had not only to be an aesthetical theory with spectacular mathematics, but what were the measurable consequences of it? That was a claim to any theory, of which he never lost sight. Besides attending the lectures De Sitter discussed the theory with Lorentz and corresponded with him during these months in 1916. An important subject was the rotation of the sun and the description of the gravitational field inside and outside the sun. Rotation was a tricky point in the development of the new theory. In a letter of 26 April De Sitter hoped that Lorentz could find a *very stupid mistake* in his calculations. Along with the letter he sent his calculations, which led, in his own words, to a number of *absurd* conclusions. That night a short letter followed with a further reflection on the question what it really meant that the sun is rotating. Is it that a point on the equator and a point at an infinite distance ($r = \infty$, a star) have different angular velocities in each co-ordinate system? Is it a rigid body with molecular forces?

But that goes beyond my powers.

The next day he wrote that his equations were wrong after all, he had forgotten some terms. He ended his letter with the sentence:

I think that few people make so many errors as I do—improving them usually takes me at least as much time as the calculations themselves [AL].

But all his efforts concerning the gravitational field of the sun led within a very short time to a beautiful result. Already on 24 June De Sitter presented an article to the Academy about 'The movements of the planets and the moon according to Einstein's theory' (De Sitter 1916a). In this article De Sitter started from the field equations of Einstein from 1915 (Einstein 1915) and for the field round the sun he

Fig. 9.4 Johannes Droste. Photograph from the Liber Amicorum for Hendrik van de Sande Bakhuyzen

based himself on the work of Droste. Again he found the correct value for the perihelion shift of Mercury, but he arrived at the result, by introducing a few acceptable approximations, in a simpler manner than Einstein had done.

He even ventured to make calculations on the moon, moving in the combined gravitational fields of the sun and the earth. For the moon he found a secular[2] movement of the perigee (the point nearest to the earth) of $1''.91$ per century. It must have given De Sitter much joy to have been able to find a simplified solution of the derivation of Mercury's perihelion shift, and also to have found a result for the moon with the help of the perturbation theory, although at the time it was not clear it could be measured.

---

[2]*Secular* means a permanent movement, not a periodical one.

De Sitter had sent the metric tensor he had used by letter to Einstein, who answered on 22 June: that he was curious about the moon movement, and that he was delighted De Sitter found so much pleasure in general relativity [*CPAE*, 8A, 227]. Einstein made use of his new acquaintance in The Netherlands. He asked De Sitter's advice on the succession of Karl Schwarzschild,[3] director of the Potsdam Observatory, which he gave elaborately. He named Karl F. Küstner (1856–1936), who would receive an honorary doctorate of Leiden University in 1928, and Hertzsprung, whom he himself would appoint deputy-director of Leiden Observatory two years later. In the same letter he wrote to Einstein that he was busy with an essay on the new gravitation theory and its astronomical consequences for the *Monthly Notices of the Royal Astronomical Society*.

> In England your theory seems to be still virtually unknown [*CPAE*, 8A, 243].

In 1916 and 1917 Einstein and De Sitter kept up a regular correspondence about a number of fundamental points in Einstein's theory. It led to two controversies between the men, who still became good friends in this period. A lot of their correspondence also dealt with their first cosmological models, describing for the first time the whole universe.

De Sitter Meets Einstein—First Controversy: The Distant Masses

Einstein rejected Newton's absolute space, but he thought he needed faraway masses to explain inertia. De Sitter presented an article to the Academy about the relativity of rotation in Einstein's theory on 30 September 1916 (De Sitter 1917a). About the *distant masses* he took up the gauntlet against Einstein with the remark:

> It seems to me that Einstein made an error here.

The necessity for the existence of distant masses followed from Einstein's wish to make the gravity field at infinity zero in each co-ordinate system. In a precise analysis of a rotation experiment De Sitter showed that the condition for the gravitational field to be zero at infinity is part of the concept of an absolute space, which has no basis in a relativity theory. At the end of the article he analysed the difference between translation and rotation. Rotation is relative in Einstein's theory, but not equivalent to translation. A transformation to let disappear a translation, does not alter the $g_{\mu\nu}$ and thus has no influence on the gravitation. In the case of a rotation it is also possible to choose a co-ordinate system in which an object does not rotate, but the rotation does not disappear. Then in Newton's theory inertial forces in absolute space appear. In Einstein's theory, which knows no difference

---

[3]Karl Schwarzschild (1873–1916) gave in 1916 the first solution of the Einstein equations, for the field outside a spherical mass.

between inertia and gravitation and does not know absolute space, the classical inertial forces are of the same kind as gravitational forces.

Mach's Principle

Ernst Mach (1838–1916), physicist and philosopher, wrote his famous work *Die Mechanik* in 1883, about the development of mechanics from Archimedes until Newton. In this book he uttered criticism on Newton's concepts of mass, space and movement. He did not accept concepts which did not originate from the tangible experience. He rejected Newton's mass definition as density times volume as a circular reasoning, because for the definition of density you needed again mass. His criticism on Newton's absolute space, absolute time and abslote moving were more important. He criticised Newton's *bucket experiment*, in which he tried to prove the existence of absolute space by the appearing centrifugal forces.

By starting to rotate a bucket with water round the vertical axis the water starts to rotate as well and creeps up the side of the bucket, by which a concave surface appears. If you suddenly stop the bucket, the water stays rotating for some time with a concave surface. So the relative movement between bucket and water is not the origin of the concave surface. For Newton there had to be an absolute space, relative to which the water was rotating. However, Mach was of the opinion that the concave surface originated by the rotation relative to the earth and the fixed stars far away.

A lot has been thought and written about the bucket experiment over the years. For instance what will happen if the wall of the bucket is given a thickness of many kilometres? Or if around the bucket, far away from it, would be a shell of mass instead of the fixed stars?

** What will happen if the shell is at rest and the bucket with water is rotating?

** And what will happen if the shell is rotating and the bucket with water is at rest?

According to Newton a concave surface appeared in the first case and a flat surface in the second (relative to absolute space).

According to Einstein the surface would be concave in both cases (relativity).

Einstein stayed a faithful supporter of the idea of the indispensable distant masses for a long time. Later he had to abandon the idea. But he stayed an admirer of Mach and acknowledged that he had been important for his scientific development.

De Sitter and Einstein met for the first time in September 1916 (Van Delft 2006). Einstein had come to Leiden at the invitation of Ehrenfest. It had taken Einstein a lot of trouble to get all his travelling documents in order. World War I was going on, but by the neutral status of The Netherlands travelling was not impossible.

Fig. 9.5 Ernst Mach

In Leiden they started their two years' discussion about the fundamentals of cosmology. In his notebook *Studies S12* De Sitter wrote a report of a discussion with Einstein, Ehrenfest and Nordström[4] about the relativity of the rotation on 28 September 1916. About this and/or other meetings with Einstein at the De Sitters his wife would write later in her book about her husband (De Sitter-Suermondt 1948):

> Downstairs I heard a ceaseless walking back and forth and heated arguing.

Einstein wished the hypothesis of the *closedness of the world*. He maintained that there had to be masses at infinity (at a mathematical finite distance, but farther away than all observable objects), so that the values of the metric tensor $g_{\mu\nu}$ are the same in every co-ordinate system. However, that could not be true for all systems. Einstein was prepared to give up the complete freedom of transformations and to restrict them to one time and three space co-ordinates. If this could not be the case, which De Sitter and Ehrenfest thought, then the hypothesis of the closedness was untrue. If this could be the case, then after all De Sitter considered it contrary to the

---

[4]The Finnish theoretical physicist Gunnar Nordström (1881–1923) belonged to the first group of theoretical physicists an astronomers working on the new theory immediately after its publication.

Principle of Relativity. The next day Einstein came with the following degenerate values for the elements of the metric tensor:

$$
g_{\mu\nu} = \begin{pmatrix} 0 & 0 & 0 & \infty \\ 0 & 0 & 0 & \infty \\ 0 & 0 & 0 & \infty \\ \infty & \infty & \infty & \infty^2 \end{pmatrix}
$$

In this case the speed of light at infinity would become infinite. In a long footnote in his article about rotation (De Sitter 1917a), which he presented to the Academy a few days after his discussions with Einstein, De Sitter mentioned Einstein's idea. But in his opinion that would make space and time at infinity again absolute and Newtonian. That went too far according to De Sitter, because it would make the world finite. The first controversy, about the distant masses, Mach's principle, was born. De Sitter would write about the infinite values proposed by Einstein that the cure was worse that the disease.

The England Connection

In that same year 1916 the English astronomer Arthur S. Eddington wrote to De Sitter that he and others were very interested in Einstein's new theory, but that nobody had yet been able to read the original articles [*CPAE*, 8A, 243].[5]

Arthur Stanley Edington (1882–1944)

Eddington studied at Trinity College in Cambridge, worked at the Royal Greenwich Observatory and was appointed professor of astronomy in Cambridge in 1913. He developed a model of the interior of the sun and he realized as one of the first English scientists the full importance of the general theory of relativity. His contacts with De Sitter and De Sitter's articles in the English language played an important role in making the theory known in England. With others Eddington organized the expedition to the island of Principe (200 km west of Equatorial-Guinea) in 1919 in order to measure the bending of light. The success of this expedition made Einstein world famous.

At an astronomy dinner he recited a poem, from which the following four-line stanza:

> Oh, leave the wise our measures to collate
> One thing at least is certain, light has weight
> One thing is certain, and the rest debate
> Light rays, when near the sun, do not go straight.
> (Vibert Douglas 1956)

---

[5]See (Röhle 2007), References of Chap. 8.

Fig. 9.6 De Sitter and Eddington. Photograph from the Leiden Observatory archive

The cause of this was World War I, which had brought the scientific contacts between Germany (one of the central powers, with Austria-Hungary, the Ottoman Empire and Bulgaria) and the allied countries virtually to a complete standstill. By its neutrality The Netherlands could function as link. De Sitter had been already an associate member of the Royal Astronomical Society since 1909. And by his other work and publications he was a known and valued astronomer in England. He had already published about several aspects of the old gravitation theory and the special theory of relativity. De Sitter immediately reacted positively. He was eagerly prepared to work on it and communicate on the new gravitation theory to inform the English scientific community. Actually, Eddington had been interested in gravitation for a long time. He was a participant of the eclipse expedition to Brazil in 1912, when one already tried to measure the apparent shift of a star in the sky caused by the gravitation of the sun. The expedition ended in a failure by heavy rains on 10 October, the day of the eclipse (Vibert Douglas 1956).

Supported by De Sitter's work Eddington became the authority in England on the general theory of relativity. On the request of the Physical Society of London Eddington published his *Report on the Relativity Theory of Gravitation* in 1918, in which he gave a complete account of the theory in an excellent manner (Eddington 1918). Actually it was the first textbook on the new theory. A favourable review appeared in *The Observatory* of April 1919, signed S., calling Eddington's work one of the masterpieces of current scientific literature. The author must have been De Sitter, who had also corrected the printer's proofs.

'Space, Time and Gravitation'

De Sitter's first publication for the English scientists appeared in *The Observatory* in October 1916. He wrote it at his holiday address in Loenen in July (De Sitter 1916b). The village of Loenen lies on the river Vecht, in the Dutch province of Utrecht. There are hardly any formulae in the article. Still it deals with all the fundamental aspects of the general theory of relativity. He did justice to Poincaré, who had already rejected the Newtonian concept of a flat 3-dimensional Euclidean space and a completely independent time many years earlier. To show the free choice of a co-ordinate system he made a connection with his other work, in celestial mechanics. In his work on periodic orbits in more-body problems the astronomer Darwin had made use of rotating co-ordinate systems in his calculations. The equality of each co-ordinate system means that a system is chosen for convenience, not because a different system is wrong. De Sitter put the emphasis on the notions world-line and observation. A world-line is a sequence of place-time events (x, y, z, t), for instance the moving of a short light vibration. An observation is always an intersection of two world-lines, for instance one of a light vibration and one of a telescope lens. We do not know how the world lines lie between the intersections. At transformations between two co-ordinate systems, no new world lines appear, no world lines vanish and their order stays unaltered.

Then De Sitter introduces gravitation. Let the acceleration of an object in a gravitational field be given by $a = d^2x/dt^2$ in co-ordinate system S (the second derivative of the spacial co-ordinate to time, leaving out y and z for easiness; the first derivative yields the velocity $v = dx/dt$). By changing to a different co-ordinate

Fig. 9.7 The big five. From left to right: Einstein, Eddington, Ehrenfest, Lorentz and De Sitter. The photograph was taken in Leiden by De Sitter's son Ulbo on 26 September 1923. Photograph from the family archive of granddaughter Tjada van den Eelaart-de Sitter

system S′ the acceleration can be transformed away, giving $a' = d^2x'/dt'^2 = 0$. In this transformation from x and t to x′ and t′ the quantities of the metric $g_{\mu\nu}$ of space appear. Thus, in view of the Equivalence Principle the metric must have a relation with the phenomenon of gravity. In other words, gravity is not a force in the classical meaning of the word. For the determination of the $g_{\mu\nu}$ Einstein formulated his equations. These equations do not start from the old gravitational theory of Newton and do not contain new constants besides the speed of light c and Newton's old constant of gravity G. The equations of Newton can be obtained in an approximation. If in this approximation the next term is also taken into the calculations, it yields in the field of the sun the missing perihelion shift of Mercury. In this publication De Sitter explained the fundamental ideas of the new theory without mathematical means, in a manner he himself understood them. In the meantime he stood so far above the theory that he could describe the essence of it without complicated calculations.

'On Einstein's Theory of Gravitation, and Its Astronomical Consequences. First Paper': Derivation of the Theory

Already in 1916 De Sitter's first famous article in a series of three appeared in the *Monthly Notices of the Royal Astronomical Society*, dated Domburg,[6] August 1916 (De Sitter 1916c). He derived the new theory from first principles and mentioned that a lot of results in his article originated completely or partly from a series of lectures at the beginning of 1916. He did not mention Lorentz's Monday morning lectures. He did mention his discussions with Lorentz, Ehrenfest and Droste explicitly. In the second half of the article he analysed the gravitational field of the sun, the redshift of light rays originating from the sun, the very small influence on the gravitational field resulting from the rotation of the sun, and the orbit of a small planet round the sun. From these last calculations he derived changes in the Keplerian laws and of course again the extra perihelion shift of Mercury.

'On Einstein's Theory of Gravitation, and Its Astronomical Consequences. Second Paper': Practical Calculations and the First Controversy

De Sitter wrote the second article for the *Monthly Notices* in Leiden, in September and October 1916 (De Sitter 1917b). Besides the colleagues and friends, already cited in the first paper, he mentioned the great privilege having had a number of discussions with Einstein in Leiden. De Sitter started the article with a description of the gravitational field of a number n of bodies, based on the work of Droste. His aim was to calculate the movements of one body under the influence of the other bodies. In a co-ordinate system with the origin in the centre of the sun he calculated

---

[6]De Sitter stayed probably in the house of his sister Wobbine in Domburg, at the seaside in the province of Zeeland.

the effects on the orbit of the moon. These were the sum of three parts. The first was the two-body problem moon-earth alone, from which followed a rotation of the perigee of 0″.06 per century. Theoretically interesting, but very small. The second part treated the influence of the sun and the earth, as if the other was not present. And the third part treated the *interference* of the two fields on the moon. The interference effects produced only extremely small oscillations in the shift of the perigee, which could be neglected.

The effect of the superposition of the earth and the sun without interference yielded a perigee shift of 1″.91 per century. To which the value of the two-body problem had to be added. Within the accuracy of the existing measurements of the moon orbit these values could not be used as proof, pro or con, of Einstein's theory. Later it was called the De Sitter effect. A definite confirmation was obtained in 2007 by measurements of the artificial satellite Gravity Probe B.

After these practical calculations De Sitter laid the values of the metric tensor at infinity, presented to him by Einstein, on the analytical grill, continuing the discussion on their first controversy. He showed that the distant masses were necessary to create the degenerate values of the metric at very large distances. But De Sitter argued that an explanation with the help masses that had never been observed was equal to no explanation at all or to acknowledging our ignorance. De Sitter argued further. The hypothesis of the degenerate metric and the distant masses is not an essential part of Einstein's theory. We do not have to accept the hypothesis and may hope that later a satisfying explanation of the inertia will be found, he wrote. De Sitter ended by praising Einstein's new theory. It encompassed the old theory, but explained also the perihelion shift of Mercury, and a number of additional phenomena, like the redshift of light rays from the sun and the bending of light rays near a gravitational field.

De Sitter wrote an elaborate letter to Einstein op 1 November 1916, in which he expounded again his objections against and doubts about the distant masses, just as he had done in the last paragraphs of his second article for the Astronomical Society. On the same day De Sitter sent a printed copy, probably of his article in *The Observatory* [CPAE, 8A, 272]. De Sitter thought that the distant masses would end like the ether wind. He asked himself if the explanation of inertia perhaps had to be found in the infinite small instead of in the infinite large. He preferred no explanation at all above that of Einstein, which actually was no explanation. Einstein answered by return [CPAE, 8A, 273]. He eased off a bit concerning the question of the values of the metric at infinity and called it a matter of taste, which would never get a scientific meaning. Apart from that Einstein was very grateful to De Sitter to have *bridged the abyss of blinding*, meaning the First World War, for his contacts with Eddington and his articles in the English language [CPAE, 8A, 290].

Einstein Has New Ideas, End of the First Controversy

At the start of 1917 Einstein would come to the conclusion that there was an alternative way to describe the universe, without the distant masses at infinity. He wrote on 2 February 1917, that he had abandoned the righteously by De Sitter disputed degeneracy of the values of $g_{\mu\nu}$. And he asked what De Sitter would find of a somewhat fantastic new idea of his [*CPAE*, 8A, 293]. By choosing a spatially closed universe, there was no infinity anymore and consequently no boundary conditions there. The distant masses were not necessary anymore. The universe finite, curved in itself and closed, but boundless. For comparison one may think of an 1-dimensional world, curved into a circle: finite but boundless. In a later added note of May 1917 about his discussions in September 1916 in his notebook *Studies S12* De Sitter wrote that he had no objection against finiteness without bounds [*ASL*, *Studies S12*]. Einstein would write to Ehrenfest two days later that his new idea implied a danger of him being locked up in a lunatic asylum. Hopefully they hadn't one in Leiden, so that he could visit Ehrenfest safely [*CPAE*, 8A, 294]. On 14 February he wrote again to Eherenfest, who would perhaps consider his way out *adventurous* [*CPAE*, 8A, 298]. He also asked after De Sitter's health which he was worried about. A short time later De Sitter would be admitted to a sanatorium in Doorn for a period of three months. Ehrenfest had to show Einstein's new work, which he sent along, to Lorentz and De Sitter. Einstein presented his new ideas to the Prussian Academy in February 1917 (Einstein 1917).

The Cosmological Constant Λ

Einstein had changed to the idea of a closed universe (Realdi and Peruzzi 2009). In order to get some insight, we consider a 1-dimensional space or *line world*. If it is a straight line, the world would extend to infinity. But that was the origin of all the discussion. Einstein's mad idea was to curve the line world like a circle and to close it. In a figure we can draw this closed universe as a circle round the origin in a 2-dimensional plane with an xy-co-ordinate system. This universe has thus a well-defined radius R. If we follow this universe in time, we can use the z-axis as time axis. This circular universe draws a cylinder through a 3-dimensional space. This is why Einstein's new universe was called the cylinder universe. With two extra spatial dimensions Einstein needed an extra dimension to curve his space and close it, a dimension we cannot imagine. Together with the time axis we can give a complete mathematical description in five dimensions. It was hardly imaginable, but it had a great advantage: Einstein could dispense with the distant masses.

But in his analyses Einstein had encountered a major problem. He had substituted the metric tensor $g_{\mu\nu}$, of which the 16 elements described his cylinder universe. But he could in no manner arrive at a solution of his equations $G_{\mu\nu} = -\kappa$ $(T_{\mu\nu} - \frac{1}{2}Tg_{\mu\nu})$, to which he put the additional condition that it had to represent a static universe. In the solution he did get the universe was not static, it would collapse as a result of its own gravitation. But Einstein saw a way out. He could add the term $-\Lambda g_{\mu\nu}$ to his equation. The equations stayed covariant and had the new form: $G_{\mu\nu} - \Lambda g_{\mu\nu} = -\kappa (T_{\mu\nu} - \frac{1}{2}Tg_{\mu\nu})$. So suddenly a new constant appeared in the equations, which was soon called the cosmological constant. In the solution this constant produced a certain counterforce, to prevent the universe collapsing under the influence of its own mass. A new constant was of course a disadvantage, because the theory had, as De Sitter had remarked, yielded beautiful results without the introduction of new quantities or constants. It is of course possible by the introduction of new quantities in a theory to explain new phenomena, but the aim of a theory is always to explain a maximum number of phenomena with a minimum of quantities and constants. This aim is in agreement with the principle of the four-teenth century philosopher William of Ockham, called Ockham's razor. A second possible disadvantage was that the value of Λ had to be very small. The reason was that in the approximation to a flat space the normal Newtonian law of gravity had to appear, without too much disturbance of the new cosmological constant. But the advantage was that the addition of the cosmological term did lead to a static solution, something which Einstein considered absolutely necessary. In the new equations he could insert the chosen metric and arrive at a static solution. As a result of his calculations he arrived at a few equalities: $R = 1/\sqrt{\Lambda}$ and $\rho = 2\Lambda/\kappa$. Einstein had the achieved formulae for the radius of the universe and its average mass density ρ. From the data then available to Einstein, he made a guess for the density: $\rho \approx 10^{-22}$ g/cm^3. That gave him from the second equation a value of Λ, which led with the help of the first formula to an estimate of the radius of the universe. This value was 10,000,000 light years, being much larger than the diameter of the visible space in those days of 10,000 light years. The density was determined with data from our own galaxy, in those days the visible universe, and yielded a value which was too large. The real value has to be calculated not from the density of one galaxy, but as an average over all galaxies, with enormous voids in between. This would lead to a much larger value of R. But this was not known to Einstein at the time. Einstein wrote about his *castle in the air* to De Sitter [*CPAE*, 8A, 311]. Among other subjects he wrote about the stars at *the other side* of the universe and about the negative parallaxes they had to produce (in a closed universe you could in principle see a star in two directions). He was in doubt and took himself not quite seriously:

> Now enough about it, otherwise you will laugh at me.

Einstein was glad he had been able to finish all the calculations without encountering contradictions. But added:

> If it agrees with the real world, is a different question.

De Sitter, in the meantime again ill and bedridden, answered on 15 March with a postcard, writing that he admired the result, but was glad that Einstein did not wish to impose it to reality [*CPAE*, 8A, 312].

The De Sitter Universe and the Second Controversy

De Sitter wrote a letter from Leiden five days later, which upset Einstein very much [*CPAE*, 8A, 313]. De Sitter had found his own solution of the Einstein equations with the cosmological constant. But: *without mass*, two words he underlined. The right-hand side of the equations became zero: $G_{\mu\nu} - \Lambda\lambda_{\mu\nu} = 0$. The right-hand side contained mass, energy and impulse. So without mass it became zero. The metric tensor chosen by De Sitter yielded only zero's at infinity and was as a consequence invariant for all transformations there. In our neighbourhood the metric tensor transformed to that of the old relativity theory. For the radius of his universe De Sitter found: $\Lambda = 3/R^2$. And for the mass density of course $\rho = 0$. De Sitter called Einstein's universe Model A and his own Model B. His own Model B could also be considered as a finite universe, for which the simple geometrical picture of a cylinder, was replaced by a hyperboloid, a rotated hyperbola. De Sitter asked Einstein for comments and hoped that Einstein's health was better. At that time both had problems with their health. De Sitter presented his work to the Academy on 31 March (De Sitter 1917c). In a footnote he mentioned that Ehrenfest had put him on the track to his new solution a few months earlier. Einstein answered by return [*CPAE*, 8A, 317]. He gave a number of objections of a technical-mathematical kind, from which he concluded that De Sitter's Model B could not describe a physical possibility. For Einstein it was unsatisfactory that there could be a universe without mass. The metric field should have been determined by the masses. So without mass it actually could not exist. The second controversy was born. De Sitter, in the meantime in sanatorium Dennenoord in Doorn, wrote back on 1 April [*CPAE*, 8A, 321]. He described his illness as not at all severe, only long-lasting and boring. He criticised the constant mass density (over space), filling the universe in Model A, whereas it appeared from all observations that this density was extraordinarily inhomogeneous. Einstein was of the opinion that he could insert the average of the existing inhomogeneous distribution of the masses in his solution; and that it was not unobservable supernatural mass, as De Sitter had suggested [*CPAE*, 8A, 351]. In July Einstein came with the idea that on the equator of De Sitter's universe, the plane at a distance $r = \pi R/2$, masses could accumulate based on the fact that in that plane the speed of a clock would become zero [*CPAE*, 8A, 363]. In that plane there would then be a singularity. In this way Einstein thought to have found a *bridge* between the models: B with all masses concentrated on the equator and A with all masses distributed evenly. But De Sitter would not hear of it.

There would still again be distant masses.

In De Sitter's opinion it would be *materia ex machina* to save Mach's dogma. Moreover, that mass would probably have to be infinitely large [*CPAE*, 8A, 370]. The rest of the scientific scene in Leiden, which stayed a scientific centre for the new theory, was of course curious after De Sitter's results and his discussions with Einstein. To bring everyone up to date De Sitter lectured on gravitation in the Ehrenfest Colloquium at the end of October [*CPAE*, 8A, 393].

But Einstein did not give in. He published an article about the fundamental points of his general theory of relativity: the Principle of Relativity, the Principle of Equivalence and Mach's Principle. He named Mach's Principle now explicitly, and not implicitly in the Principle of Relativity, to make clear that inertia had to be traced back to the interaction between the masses (Einstein 1918a). Einstein also published a critical consideration about the De Sitter universe, in which he had written down his idea that De Sitter's universe was not massless after all, but a universe in which all mass was concentrated in the plane r = πR/2 (Einstein 1918b). De Sitter answered with a letter, in which he wrote not (yet) to know if there was mass in his model or not [CPAE, 8B, 501]. Einstein was supported in his opinion by the German mathematician Hermann K.H. Weyl (1885–1955), who thought to have given a proof of Einstein's supposition with a fluid band of mass round the equator [CPAE, 8B, 506]. But another German mathematician, Felix C. Klein (1849–1925) came to the aid of De Sitter. In a few letters to Einstein he showed that the singular behaviour at the equator of the De Sitter universe was only the result of a specific substitution and had nothing to do with the existence of mass there. After having written Weyl's solution to Klein in the beginning of June, Einstein answered to Klein that he was completely right and that the De Sitter universe without mass did not contain singularities. But, Einstein added, one should certainly not consider this universe as a physical possibility [CPAE, 8B, 567]. He wrote to Ehrenfest to hope that De Sitter, who recently had been appointed director of the Observatory in Leiden, would soon be well again. De Sitter's criticism had been partially true. Einstein was sorry about the discussion [CPAE, 8B, 664].

So this marked the end of Einstein's mathematical opposition to De Sitter's Model B, and also the end of the second controversy. But he was not prepared to consider it as a description of the real world. Understandable, then what was a world without mass?

'On Einstein's Theory of Gravitation, and Its Astronomical Consequences. Third Paper': Large Receding Velocities of Distant Nebulae

De Sitter's third article in his trilogy for the Royal Astronomical Society, which he had finished during his stay in the sanatorium in Doorn, appeared in November 1917 (De Sitter 1917d). In this article he presents Einstein's universe, Model A in De Sitter's words, and his own solution, Model B, without mass. The models have been discussed in the previous paragraphs. De Sitter would not be himself, if he did not try to relate the available observational data to his theoretical model. He mentioned the fact that in his model the frequency of light vibrations decreased with the distance r to the origin. That leads to a larger period T and a larger wavelength λ. If we observe these wavelengths, we see a redshift. From a number of stars a systematic redshift had indeed be observed. De Sitter pointed out that this redshift

Fig. 9.8 Felix Klein

consisted of two parts. One part comes from the fact that the light was sent from the gravitational field of the star, and a second part originated from the fact that the component g_{44} of the metric tensor decreased for increasing r. The second part a result of De Sitter's choice for the metric of Model B. In Einstein's Model A this second part is zero. Another point, following from De Sitter's formulae, was that we could expect for the velocity of material objects at very large distances a larger part with large radial velocities. These radial velocities also gave a Doppler red- or blueshift of the spectral lines. De Sitter did not comment upon the direction of the velocities, to or from the earth. He tried to get information from the most recent observational results. The radial velocity of three nebulae, probably belonging to the farthest objects, had been determined by more than one observer. The average value was 600 km/s, receding from the earth. On the basis of only three observations, it had still little value, as De Sitter remarked. But if continued observations of distant nebulae would yield systematically positive radial velocities, it would be an argument in favour of the acceptance of Model B. If not, it would point in the direction of Model A, or Model B with a much larger R. Although there only were few observational data on radially moving nebulae, De Sitter felt he was on to something. He wrote Kapteyn for advice on the measurability of the redshifts in June 1917, but the last could not provide much clarity on the *tough nuts* De Sitter had given him *to crack* [ASL, 28-6-1917]. De Sitter was the first who, be it carefully, linked the receding nebulae to his cosmological model. One could not yet speak of an expanding universe, because De Sitter's Model B described, like Einstein's a static universe. For the first non-static models it was waiting for Friedmann[7] (1922) and Lemaître[8] (1927). From the article we take the following data about three distant objects.

> **Velocities of spiral nebulae (1917, more than one observer)**
> Andromeda (3 observers) − 311 km/s.
> N.G.C. 1068 (3 observers) + 925 km/s.
> N.G.C. 4594 (2 observers) + 1185 km/s.
> N.G.C. is the New General Calalogue of distant objects in space, started in 1880.
> The Sitter remarked:
>
> > These velocities are very large indeed, compared with the usual velocities of stars in our neighbourhood.

In his previously mentioned Report on general relativity Eddington discussed the Models A and B. He was rather critical of Einstein's model: the fact that we should see in the opposite direction an anti-sun (as a result of the closed universe), and the fact that the cosmological constant was dependent on the total mass in the universe: $\Lambda \sim 1/M^2$.[9] This seemed to him incomprehensible. He wrote:

We are sorry not to be able to recommend this rather picturesque theory of anti-suns and anti-stars.

Eddington preferred De Sitter's model:

There is no anti-sun and it provides a possible explanation of the large observed radial velocities of distant nebulae.

In spite of this critical comment about Einstein's model of the universe Eddington was one of the strongest supporters of the new theory.

---

[7]The Russian astronomer and mathematician Aleksandr A. Friedmann (1888–1925) derived the later called Friedmann equations, starting from Einstein's equations without the cosmological constant. His time dependent solutions for the universe were not noticed at the time.

[8]The Belgian astronomer and Jesuit priest Georges H.J.E. Lemaître (1894–1966) gave in 1927 a solution for an expanding universe. Later he was the first to state the possibility of a big bang. The term big bang was coined by the British astronomer Fred Hoyle in the 1950s as a deprecation, he did not believe in the expanding universe.

[9]This dependence follows directly from the formulae: $\Lambda \sim 1/R^2$, $\Lambda \sim \rho$ (both valid in Model A) and $M \sim \rho R^3$.

The Eclipse Expeditions in 1919

The Astronomer Royal of England Sir Frank Dyson realized already during the First World War the excellent possibility to measure the predicted bending of light rays by the gravitational field of the sun during the total eclipse of the sun on 29 May 1919. With a grant of £1000 and the support of Eddington preparations were made. An important advantage was that the armistice was signed in 1918, ending the First World War. Two expeditions were organized, one to Sobral in northeast Brazil, and a second one to the island of Principe, west of Africa. The bending should be, according to Einstein's theory, $1''.75$, while on the basis of the older theory it was $0''.78$ or even zero. Eddington joined the expedition to Principe, where the group arrived on 23 April 1919. During the nights they could make enough control pictures. The weather on 29 May was bad, a lot of rain and clouds, and not until the partial eclipse phase the sun appeared more or less. Sixteen photographs were taken. The sun was clearly to be seen on all of them, but by the clouds only a few stars could be seen on the last six photographs, according to Eddington's diary (Vibert Douglas 1956). The photographic plates were developed and measured during the days after the eclipse, but as a result of the bad weather Eddington had to treat his measuring results in a different way as planned and could therefore not make a preliminary communication on the results. But he was able to measure completely one of the plates and compare it to other data. That measurement gave a result in agreement with Einstein's prediction. Later Eddington would call that the most important moment in his life. The measuring of the plates, the calculations and the comparison to reference pictures took some time. At the meeting of the British Association for the Advancement of Science on 12 September 1912 in Bournemouth Eddington announced the preliminary result: between $0''.87$ and $1''.8$ s of arc. The message reached Einstein via Leiden, but he had still to wait till the final announcement of the result on 6 November at a combined meeting of the Royal Society and the Royal Astronomical Society. Dyson said:

> After a thorough study of the plates I am prepared to say that they confirm Einstein's prediction. (Pais 2005)

On that day Einstein was canonized by the international press. The predicted redshift of light rays coming from the sun had still to be observed, but the scepticism with most scientists had vanished. *Einstein* became from a proper name an appellative name: *an Einstein* stands for a person of extremely high intelligence.

Lorentz congratulated Einstein, Dyson wrote to Ejnar Hertzsprung, who had recently been appointed deputy director at Leiden Observatory under its new director De Sitter, who stayed at that time in a sanatorium in Arosa in Switzerland. De Sitter and Fokker, at that moment together in Arosa, wrote their congratulations to Einstein. Fokker wrote on 18 November, in which he said that De Sitter had received a copy of Dyson's words. De Sitter did not write until 1 December, because he was confined to bed. Fokker invited Einstein warmly, if he was tired of all the fuss, to visit them, exiled in mountain and snow. De Sitter wrote: And now still the redshift. But he had little hope that could be measured. The atoms in the sun were not astronomical clocks and could be disturbed by all kinds of causes, although Einstein had a much more positive view on the possibility [*CPAE*, 9, 208]. De Sitter wrote:

> Fokker and I try to keep each other scientifically awake.
>
> [*CPAE*, 9, 168] [*CPAE*, 9, 185].

Einstein wrote to Fokker that his joy about the propagation in England of his theory and the ensuing successful expeditions were largely due to De Sitter. Besides there were his own contributions [*CPAE*, 9, 187].

Einstein Special Professor in Leiden. De Sitter's Provisional Farewell to Cosmology

The bond between Einstein and Leiden became a formal one. Ehrenfest took the initiative to tempt Einstein to come regularly to Leiden. In 1919 he wrote a typically *Ehrenfestian* letter to Einstein, of which here follows a part:

> All of us are now in complete agreement that we must undertake to get you to Leiden. The business is extremely simple: if you just say yes to me, it will be possible—at least according to all human expectations—to arrange things extremely quickly according to your wishes ... You can spend as much time as you want in Switzerland, or elsewhere, working, giving lectures, travelling, etc., provided only that one can say *Einstein is in Leiden—in Leiden is Einstein*—...Dear, dear Einstein! Don't destroy all my dreams and hopes. Help me in my efforts by immediately sending me an answer favourable enough for me to put everything in motion directly.[10]

Einstein consented and would come to Leiden every year for a couple of weeks. He gave his inaugural lecture on 5 May 1920, titled *Ether and Relativity Theory* (Einstein 1920). Ehrenfest had suggested to him the title *Away with the ether superstition* [*CPAE*, 9, 373], but Einstein could not let this pass his lips, because of his great esteem for Lorentz. Einstein gave the old concept of the ether a new content:

> According to the general theory of relativity space has physical properties; thus in that sense there exists an ether.

By these physical properties Einstein meant the components of the metric tensor.

On the sudden demise of Ernst van de Sande Bakhuyzen in the beginning of 1918 De Sitter became director of Leiden Observatory. His own appointment procedure, the highly necessary reorganization of the Observatory, and the appointment of two (later one) deputy directors, all laborious processes, required in

---

[10]The translation is made by the American historian of science Martin J. Klein (1924–2009) in his partial biography of Ehrenfest (Klein 1970).

1918 and 1919 all his, diminishing, energy. After that De Sitter stayed nearly a year and a half in a sanatorium in Arosa in Switzerland, where he did have time to do a lot of calculations on cosmology and his new theory of Jupiter's satellites.

During the decade that followed, managerial work in the Observatory, the University and the International Astronomical Union, and a lack of relevant measurements on nebulae, prevented De Sitter to put much effort in cosmology. For cosmology he found time again in 1929.

The Blessing of Not Being the Observatory Director

It is tempting to philosophize about what would have happened if De Sitter in 1908 had been appointed director of Leiden Observatory after all, along with his appointment as a professor. Although De Sitter found in 1918 a dozed off Observatory and had to implement a complete reorganization taking a number of years, the situation in 1908 was slightly better but far from optimal. The main activity of the Observatory had not been changed since 1860, when director Kaiser built the Observatory: fundamental, positional astronomy. The new developments, photography and spectroscopy, hardly played a role. The management of the small Observatory, with ten to twenty workers, would have posed no problem for De Sitter. But it would have taken him a lot of time to set up new measuring programmes, the reduction of available, mostly old measurements into astronomical data, and the introduction of modern developments in astrophysics.

In retrospect it seems to have been good for De Sitter that his career was a two-stage rocket. The first period, between 1908 and 1918, was a very fruitful decade in theoretical astronomy. It made his name in the astronomical world. As a consequence he could, when he became director of the Observatory, by his scientific influence and authority, and by the support of Kapteyn, then the grand old man of Dutch astronomy, convince the government to finance a full-scale renovation of the Observatory. He managed to persuade the then already world-famous astrophysicist Ejnar Hertzsprung to come from Potsdam to Leiden. Comparable results would probably not have been reached by De Sitter in 1908. And moreover he would not have been able to develop fully his new theory of Jupiter's satellites and his cosmological work. Or all of this only partly. De Sitter saw the dangers already in 1908. He wrote a letter about it to Gill [RGS, DOG/159, 30-11-1908]. In that letter he explained the events round his appointment and finished with some detachment, understatement and self-knowledge:

> I confess I have not been entirely averse of the idea of taking up the directorate and facing all the difficulties connected with it—though they are very great—and trying to make the Leiden Observatory work as good as it can. But I clearly saw that in that case it would be impossible for me to do any work myself, at least of a theoretical kind, until I should have trained a staff which worked after my ideas.

References

De Sitter, W. (1916a). 'De planetenbeweging en de beweging van de maan volgens de theorie van Einstein' (Movement of a the planet and the moon according to Einstein's theory); *NAW2*, 25, pp. 232–245, 1916.

De Sitter, W. (1916b). 'Space, Time and Gravitation'; *The Observatory*, 39, pp. 412–419, 1916.

De Sitter, W., Assoc. R.A.S. (1916c). 'On Einstein's Theory of Gravitation, and its Astronomical Consequences, First Paper'; *MNRAS*, LXXVI, 9, Supplementary Number, pp. 699–728, 1916.

De Sitter, W. (1917a). 'On the relativity of rotation in Einstein's theory'; *NAW2*, 19 I, pp. 527–532, 1917.

De Sitter, W., Assoc. R.A.S. (1917b). 'On Einstein's Theory of Gravitation, and its Astronomical Consequences, Second Paper'; *MNRAS*, LXXVII, 2, pp. 155–184, 1917.

De Sitter, W. (1917c). 'Over de relativiteit der traagheid: Beschouwingen naar aanleiding van Einstein's laatste hypothese' ('On the relativity of inertia: Considerations in response to Einstein's latest hypothesis'); *NAW2*, 25 II, pp. 1268–1287, 1916–1917.

De Sitter, W., Assoc. R.A.S. (1917d). 'On Einstein's Theory of Gravitation, and its Astronomical Consequences, Third Paper'; *MNRAS*, LXXVIII, 1, pp. 3–28, 1917.

De Sitter-Suermondt, E. (1948). *Een menschenleven*. See References of chapter 3.

Eddington, A.S. (1918). *Report on the Relativity Theory of Gravitation*, Fleetway Press, London, (1918, 1920). Republication: Dover Publications, Inc., Mineola, N.Y., USA, 2006.

Ehrenfest, P. (1909). 'Gleichförmige Rotation starrer Körper und Relativitätstheorie'; *PZ*, 10. Jahrgang, 23, p. 918, 1909.

Einstein, A. (1915). 'Die Feldgleichungen der Gravitation'; *Sitzungsberichte Preussische Akademie Berlin*, pp. from 844, 25 November 1915.

Einstein, A., (1917). 'Kosmologische Betrachtungen zur allgemeinen Relativitätstheorie'; *Könichlich-Preussische Akademie der Wissenschaften. Sitzungsberichte*, pp. 142–145, 1917.

Einstein, A. (1918a). 'Prinzipielles zur allgemeinen Relativitätstheorie'; *AP*, 55, pp. 241–244, 1918.

Einstein, A. (1918b). 'Kritisches zu einer von Hrn. De Sitter gegebenen Lösung der Gravitationsgleichungen'; *Könichlich-Preussische Akademie der Wissenschaften. Sitzungsberichte*, pp. 270–272, 1918.

Einstein, A. (1920). *Äther und Relativitätstheorie*, inaugural lecture, Julius Springer, Berlin, 1920.

Janssen, M. and Lehner, C., editors (2014). *The Cambridge Companion to Einstein*, Cambridge University Press, 2014.

Klein, M.J. (1970). *Paul Ehrenfest, Volume 1: The making of a Theoretical Physicist*; North-Holland Publishing Company, Amsterdam, London, American Elsevier Publishing Company, Inc., New York, 1970. (Note: Volume 2 did not appear.)

Kox, A.J. (1993). 'Leiden and general relativity'; *Classical and Quantum Gravity*, 10, pp. 187–191, 1993.

Pais, A. (1982, 2005). *Subtle is the Lord*, Oxford University Press, Oxford, 1982, 2005. This biography is an excellent source about life and work of Einstein, written by a physicist who knew him well.

Realdi, M. and Peruzzi, G. (2009). 'Einstein, de Sitter and the beginning of relativistic cosmology in 1917'; *General Relativity and Gravitation*, 41, pp. 225–247, 2009.

Van Delft, D. (2006). 'Albert Einstein in Leiden'; *Physics Today*, pp. 57–62, April 2006.

Vibert Douglas, A. (1956). *The Life of Arthur Stanley Eddington*, Thomas Nelson and Sons Ltd., Edinburgh, 1956.

Chapter 10
Director of the Observatory and Stay in Arosa

> *Here we have the most beautiful weather in the world, and splendid clear skies. This would be an excellent location for an observatory.*
>
> Willem de Sitter from Arosa
> [ASL, 20-11-1919 to Hertzsprung]

Messy State of the Observatory

The situation at the Observatory in Leiden was far from ideal. Little management, few observations, few publications. The 'young' Van de Sande Bakhuyzen would turn 70 in 1918, the regular age to retire. Some activity arose round the Observatory at the end of 1917. Mr. Rh. Feith Esq., member of the Board of Governors of the University, said in a meeting of the Board, that he had heard from reliable sources that part of the staff of the Observatory were less qualified [AUL, BG minutes]. The case was held over to the next meeting, but it was decided not to enter into proposals of the director in the meantime. In the next meeting the *situations* were discussed at the hand of a letter of Jan van der Bilt, astronomer in Utrecht, to the Board of Governors. De Sitter and Kapteyn also discussed the necessity to intervene in February 1918, but they decided to wait until De Sitter would be completely recovered. De Sitter had already missed the jubilee on the occasion of Kapteyn's 40-years' professorship on 20 February.

> **Jan van der Bilt (1876–1962)**
> After a short career in the navy Van der Bilt started his studies at the University of Utrecht in 1903. He became a student of Nijland and defended his dissertation in 1916, at the age of forty, after which he became a lecturer in 1920. He was secretary of the Eclipse Committee of the Academy from 1924 until 1946, and did a lot for the popularization of astronomy in the Dutch Society for Meteorology and Astronomy, of which he became an honorary member.

© Springer Nature Switzerland AG 2018
J. Guichelaar, *Willem de Sitter*, Springer Biographies,
https://doi.org/10.1007/978-3-319-98337-0_10

Fig. 10.1 Jan van der Bilt. Photograph from the Liber Amicorum for Hendrik van de Sande Bakhuyzen

Van der Bilt had worked a couple of months at the Observatory in Leiden and had been severely shocked by the state of things there. After a conversation with governor Feith Van der Bilt wrote the above-mentioned letter [AUL, BG, 1-1-1918]. On New Year's Day 1918 he had sat down properly to write his letter. There were serious faults: too low salaries and it was unclear how long Van de Sande Bakhuyzen would stay in office, in connection with his pension rights. Moreover, the photographic telescope was not used and was more or less promised to J. Voûte, who planned to found an observatory in the Dutch East Indies.[1]

---

[1]See Chap. 12.

Van der Bilt's proposals were the following. Let Kapteyn, Nijland and De Sitter decide who will be the successor in Leiden and let him co-operate as deputy director with Van de Sande Bakhuyzen for a year. Don't put any more money in the *ancien régime*. It is certainly possible that this letter was inspired by purely honourable worries about the languishing Observatory. But perhaps Nijland had given Van der Bilt a little boost to write the letter with the proposals, as he had formulated them, with an eye on a possible directorship in Leiden for the excellent organizer Van der Bilt.

Fig. 10.2 Albertus A. Nijland. Photograph from the Liber Amicorum for Hendrik van de Sande Bakhuyzen

The Demise of Ernst van de Sande Bakhuyzen

But the case took a turn by the sudden death of Van de Sande Bakhuyzen on 3 March [ASL, *Reorganization file De Sitter*] (De Sitter 2000; Baneke 2005). The situation of the Observatory became public. Not speaking ill of the dead was only partially granted to him. Lorentz, in his function of chairman of the Department of Mathematics and Physics of the Academy, wrote to Kapteyn requesting information on Van de Sande Bakhuyzen for his obituary speech at the meeting of 23 March. Kapteyn wrote in his answer that he had little positive to mention. First, the observations of the Venus Expedition from 1874 (44 years in the past!) had probably not yet been reduced. Second, concerning the meridian circle, Kaiser had already been clear that he had too little manpower to reduce everything. In order to solve similar problems at the Cape, Gill once had slowed down the observations in South Africa. Kapteyn, when still considering seriously in 1908 the request of the older brother Van de Sande Bakhuyzen to succeed him, had let both brothers know that in that case he would completely stop the fundamental star position measurements. Kaiser's successors had not dared to cut the knot, resulting in continuously growing arrears. A lot of reduction work would be published extremely late. Kapteyn's own measurements at the Observatory from 1875 to 1877 had not yet been published either. The Observatory had not met the expectations. The young Van de Sande Bakhuyzen had not been lacking in enthusiasm and energy, but he had been too conservative and too little bold. And as a third point Kapteyn wrote: Van de Sande Bakhuyzen was reserved towards the modern stellar astronomy. Once he had even confessed to Kapteyn:

> I have been foolish to be fixated on one point: the determination with the utmost accuracy of certain astronomical data. Therefore I have lost the aim of the real cause out of sight.

Kapteyn wrote his letter confidentially, but if Lorentz would decide to inform others, including the older brother, that would be fine with him, though he was reluctant [AL, 19-3-1918]. How different this all sounded ten years earlier, when Kapteyn (and Lorentz) recommended the young brother with his excellent and impeccable work and with the words that under his guidance the good name of the Observatory would be maintained.

De Sitter Interim Director

The Board of Directors reacted quickly, and on 9 March the Minister of the Interior already wrote that De Sitter could be put in charge of the supervision of the Observatory [AUL, BG, 9-3-1918]. On 13 March the faculty established a committee for the succession, with the members: Lorentz, Kamerlingh Onnes, De Sitter and Kuenen [AUL, Faculty papers, *Folder Vacancy*, 13-3-1918].

Fig. 10.3 Heike Kamerlingh Onnes. Photograph from the Liber Amicorum for Hendrik van de Sande Bakhuyzen

Concerning the succession Kapteyn wrote an advice at the request of the Board of Governors on 15 March. The main points in the advice were the following. De Sitter had to become director. As deputy director he proposed Pannekoek for the meridian department and for the astrophysical department Hertzsprung from Potsdam,

One of the most gifted and original astronomers.

He added that Hertzsprung was his son-in-law, so the Governors had to seek advice on him by others too. With these three men a new period of bloom would follow. But De Sitter would make his demands [AUL, BG, 15-3-1918.]. In the first few weeks after the death of Van de Sande Bakhuyzen De Sitter, being temporary director, wished to have a decisive vote in the appointment of a successor. But soon

Fig. 10.4 Johannes P. Kuenen. Photograph from the Liber Amicorum for Hendrik van de Sande Bakhuyzen

he came to the conclusion that he wished to become director himself, on the condition that Pannekoek and Hertzsprung would become deputy directors. He had ample conversations with his physician Bruining, who in the end consented, because living in the observatory was an advantage [ASL, *Reorganization file De Sitter*]. On 10 April the Board of Governors asked De Sitter's advice on his findings at the Observatory and his ideas on a reorganization [AUL, BG, 10-4-1918]. De Sitter sent an extensive report of nineteen pages within a week, in which he gave an analysis of the current situation and outlined the main points of the new direction structure, the necessary building activities and the changes in the remaining staff [AUL, BG, 17-4-1918].

Competitor Van der Bilt?

In the faculty meeting of 18 April Lorentz reported on behalf of the committee for the succession of Van de Sande Bakhuyzen. He proposed to recommend to the Governors De Sitter as director and to appoint an extraordinary professor to take on the educational task of Van de Sande Bakhuyzen. De Sitter stated to accept the directorship, provided that a number of conditions would be fulfilled by the Governors. Pannekoek was proposed as one of the deputy directors [AUL, Faculty minutes, 12-4, 18-4-1918]. According to Pannekoek De Sitter had put quite some pressure on him and that he had said yes, not wholeheartedly but still without hesitation, because giving 26 lessons a week on an hbs was not that pleasant either. Pannekoek had not the best reminiscences of the meridian department (Pannekoek et al. 1982). Kapteyn and De Sitter were invited to explain their views in the meeting of the Board of Governors on 1 May [AUL, BG, 22-4-1918]. Their opinions were clear. Kapteyn said that Pannekoek's political views would play no role anymore, and that he would dedicate himself completely to astronomy. About Hertzsprung he said that a foreigner of the *first class* was to be preferred to a fellow-countryman of the *second class*. Later during that meeting, after Kapteyn and De Sitter had left, a decision was made by majority to ask advice from Nijland in Utrecht. The questions were the following. What were Hertzsprung's qualities? Would Van der Bilt, whose qualities had been recommended to some members, give more guarantees than De Sitter as a director, or than Hertzsprung or Pannekoek as a deputy director? A minority was of the opinion that there were sufficient reasons to propose the advice of Kapteyn and De Sitter directly to the Minister [AUL, BG minutes, 1-5, 2-5-1918]. The lobby for Van der Bilt gathered some momentum. The letter to Nijland was posted on 2 May and he answered by letter of 3 May [AUL, BG, 3-5-1918]. He wrote that in the first place an organizer was needed, whose personality would be at least as valuable as his scientific merits. Supposing that De Sitter would not aspire to the directorship due to his weak health, he recommended without hesitation Van der Bilt, whose character he called as *the right man in the right place*. By mentioning De Sitter, Nijland recognized him implicitly as the best candidate. Nijland did not consider Hertzsprung a suitable candidate as a director. Neither was he positive about Pannekoek, whose appointment would be a *dangerous experiment, due to his peculiar view on matters.* Nijland sketched the in his eyes best solution: Van der Bilt director and Hertzsprung deputy director, who would operate fully independently in scientific matters. A new period of bloom was guaranteed. The Board of Governors did not follow Nijland's advice and presented to the Minister De Sitter's plan by letter of 16 May. They took quite a risk. De Sitter would only accept the appointment, if the Minister would comply with all his wishes. Thus, including the appointments of Pannekoek and Hertzsprung and a drastic rebuilding. An extensive alteration was needed for the large De Sitter family, also taking his health into account [AUL, BG, 16-5-1918].

But the problems were not yet over. There were other. Some other forces were working within the Ministry. The brother of Governor Feith, Mr. C. Feith Esq.,[2] was a high civil servant at the Ministry, in the department of education. These brothers were on friendly terms with a brother of Van der Bilt, a connection that had not to be underestimated. On 13 May there was another talk at the Ministry, where C. Feith stated that further talks were useless, because the Minister had already written to the Governors with a different plan. In this letter of 5 June the Minister wrote not to agree to the plan of De Sitter, without an argumentation. The Governors were asked after their opinion on the appointment of Van der Bilt, after which the Minister wished to receive a plan drawn up in consultation with Van der Bilt. But the Board of Governors decided not to consult Van der Bilt. They decided to ask for an audience with the Minister by president of the Board of Governors Mr. dr. N. Ch. de Gijselaar Esq. and governor Mr. dr. J. Oppenheim in order to again defend De Sitter's plan [AUL, BG minutes, 19-6-1918]. The audience took place on 25 June and the Minister promised to look into the matter again. A couple of new pieces of advice were sent to the Minister and De Sitter went to the Ministry on 7 August, where he spoke with secretary-general Mr. Jan B. Kan, who told him that there were no objections anymore against his appointment and his plans. But in the meantime the ministerial budget for 1919 was ready and changes would be very difficult to make. This was unacceptable for De Sitter. Kan said that the Governors had to write a proposal, which they did. The matter seemed in order. But the Minister, Cort van der Linden,[3] could not sign any changes in his budget, because the cabinet had resigned on 3 July.

Slowly on Course

Although a possible appointment of Van der Bilt had been averted, the appointment procedure of De Sitter still made little progress. On 3 July a general election had taken place for the Second Chamber,[4] in which all men could vote for the first time and the constituency voting system had been replaced by proportional representation. Important and financially heavy decisions were preferably postponed. The new cabinet with the Roman Catholic Prime Minister Ch. J. M. Ruijs de Beerenbrouck was sworn in by the Queen on 31 August 1918. For the first time there was a separate Minister of Education, Arts and Sciences: Dr. J. Th. de Visser, who was appointed not until 26 September 1918. Before that time education belonged under

---

[2]Mr. C. Feith became at the end of 1918 the first permanent secretary of the new Ministry of education.

[3]Mr. P.W.A. Cort van der Linden, an independent liberal, was Prime Minister and Minister of the Interior of The Netherlands during the First World War. Main achievements were the introduction of the right to vote for all men, the introduction of the proportional representation, and the equality of public schools and those based on a religious basis.

[4]Dutch: Tweede Kamer, the Dutch House of Commons.

the responsibility of the Minister of the Interior. The appointment procedure for De Sitter had ended positively after all, but he still had to wait for the appointments of the deputy directors and the remaining facilities. Minister De Visser wrote to the Board of Governors that the appointment of Pannekoek and Hertzsprung could wait till after the approval of the reorganization plans by the parliament. Their appointments as extraordinary professors could be made even later. That was disappointing news. De Sitter wrote an extensive report. Pannekoek and Hertzsprung could financially be appointed in the vacancies of Van de Sande Bakhuyzen and Wilterdink.[5] De Sitter was also afraid that a new instrument, which was a condition of Hertzsprung and was later called the *Schraffierkassette* (moving-plate camera), would be jeopardized. He proposed to the Governors to ask for an audience with the Minister. De Sitter spoke with the Minister on 3 January 1919, but the Minister stated that the budget had to be approved first.

De Sitter's friends kept in contact and sympathized with him. From South Africa Innes hoped that he would have his hands free at last [ASL, 16-7-1918]. De Sitter wrote to his friend Dyson in December that he had been appointed and that Pannekoek and Hertzsprung would come as deputy directors. He would appreciate it if Dyson would again make observations for him of Jupiter's satellites. He also mentioned his bad health and that he was not able to do much work [RGO, 8/88, 16-12-1918].

In the meantime an estimate had been made of the cost of the conversion of the Observatory building by the Government Building Department: 300,000 Dutch florins. The plan would be executed in three years. De Sitter then became really impatient. It was already 1919. Everything had taken already nearly a whole year. If possible everything had to be done within two years and the building process had to start as soon as possible.

In the end the complete renovation would be ready in 1924, the living quarters of the deputy directors were not even ready by then [AUL, BG, several letters, xx-12-1918, xx-1-1919].

Pannekoek's Appointment Thwarted

The appointments of the deputy directors were not even effectuated in 1919. During 1918 De Sitter and Pannekoek kept in close contact. Pannekoek was convinced that he would be appointed. He became a bit impatient in September and he asked De Sitter if De Gijselaar could not have a talk with Minister De Visser. Both were right-wing conservative politicians after all.

---

[5]Jan H. Wilterdink (1856–1931) worked at the Observatory as second observer from 1878, lecturer from 1901 and first observer from 1910. After the death of Van de Sande Bakhuyzen he was acting director of the Observatory for a short period, until De Sitter took over. When De Sitter officially became acting director, Wilterdink retired.

Fig. 10.5 Antonie Pannekoek. Photograph from the Liber Amicorum for Hendrik van de Sande Bakhuyzen

We possess our souls in patience.

In the same letter Pannekoek proposed, inspired by Kapteyn, that the astronomers in The Netherlands would found a society and would convene regularly. Something like the Royal Astronomical Society [ASL, 29-9-1918]. Shortly thereafter the Dutch Astronomers Society[6] was founded on 5 October 1918, under the presidency of Nijland.

In December Pannekoek tried again to bring De Sitter to some action, but he supposed it was perhaps De Sitter's illness, of which he had heard, preventing him to act more firmly [ASL, 11-12-1918]. Pannekoek did not contact his political

---

[6]Dutch: Nederlandse Astronomen Club (NAC).

friends, that would perhaps spoil things [ASL, 18-12-1918]. He had heard rumours that De Sitter had taken up the plan to go for a rigorous cure to make a definitive end to all his health troubles, for example a one year cure in Switzerland. In that case it would be of great importance that Hertzsprung and Pannekoek himself would already be living in the Observatory [ASL, 2-3-1919]. Finally in April Pannekoek wanted to know what the situation was. He was worrying. Perhaps the government had serious objections to him after all. Ministers called Bolsheviks *criminals and dangerous characters*. And small articles appeared in the papers about the time he was *literarily* active in Germany. *The Hague Post*[7] had copied an article from a German paper: *Still more roubles to The Netherlands*. In that article the suggestion was made that Pannekoek received money from Moscow [ASL, *Reorganization file De Sitter*]. You must bring in all forces, he wrote. Pannekoek complained about his work as a teacher at an hbs, 20 lessons a week[8] and correction work every day. And he had such good ideas to deduce the structure of the Milky Way, following Kapteyn's work [ASL, 15-4-1919]. A few times he asked De Sitter if he could resign already. Then, suddenly on 5 May, a letter from the Minister arrived. He refused to appoint Pannekoek, probably under pressure of the Prime Minister. Although a few futile attempts were made to turn the tide, it was soon clear that an appointment of Pannekoek did not stand a chance. De Sitter was furious with Pannekoek: What are you now, astronomer or communist? (Pannekoek et al. 1982) Pannekoek seemed to reconcile himself quickly to the inevitable. But he did advise De Sitter to resist to the utmost.

> But you must not give up, as if it is a decision of the gods, against which man is powerless. You must, so to speak, fight with nails and claws for what you worked on more than a year.
>
> [ASL, May 1919]

Pannekoek advised De Sitter also to read Kaiser's account on the building of the Observatory, in which he complained bitterly about the government [ASL, 18-5-1919]. He put the blame of his rejection on *Troelstra's idiotic November bragging* [ASL, 21-7-1919]. Pieter J. Troelstra, the leader of the Social Democratic Workers Party,[9] tried to start a socialist revolution—like in Germany and Russia— in The Netherlands in November 1918, which completely failed in a day and became known as *Troelstra's error*. But it could very well be that the failed revolution was one of the causes for Pannekoek's rejection.[10] After all Pannekoek was not very dissatisfied with not having been appointed. How much he did respect De Sitter, working under him would have inevitably led to conflicts.

> He was a *pope*, who liked to put forward his being the master. (Pannekoek et al. 1982)

---

[7]Dutch: *De Haagse Post*.

[8]In his *Reminiscences* Pannekoek said 26 lessons a week.

[9]Dutch: Sociaal-Democratische Arbeiderspartij (SDAP).

[10]Pannekoek's name as a prominent socialist was not forgotten, even after he had completely focused on astronomy from 1918 onwards. His picture was carried along on demonstrations during the so-called Carnation Revolution in Portugal in 1974.

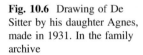
Fig. 10.6 Drawing of De Sitter by his daughter Agnes, made in 1931. In the family archive

He did not have to wait long for another astronomical position. An appointment at the University of Amsterdam was offered to him. In July the decision was made about a lecturership. The University of Amsterdam was a municipal university. So Pannekoek could not again be confronted with a *njet* of the government.

The Arrival of Hertzsprung

Ejnar Hertzsprung was born near Copenhagen in 1873. There he studied chemistry and specialized in photochemistry. This specific knowledge later helped him a lot in photoastronomy. When he was 29, in 1902, he decided to become an astronomer and started working at the Urania Observatory in Copenhagen. In 1905 and 1907 he published articles on the radiation of stars, in which he demonstrated that the bright red stars in the sky fall apart into two groups, of high and low intrinsic brightness, seeming to be connected with their size. These articles were the basis of the notions *giants* and *dwarfs*. A short time later the American astronomer Henri N. Russell (1877–1957) reached the same conclusions. The diagram, in which the intrinsic brightness (or absolute magnitude) is plotted against the surface temperature (or the spectral type) of stars is known as the Hertzsprung-Russell Diagram since then. In 1908 Hertzsprung accepted an invitation by Schwarzschild to come to Göttingen. It was the start of a long friendship. Schwarzschild said about Hertzsprung:

I am sometimes thinking. Hertzsprung is always thinking.

On the explicit wish of Schwarzschild Hertzsprung went with him to Potsdam near Berlin, where Schwarzschild had been appointed director of the Observatory. In 1912 Hertzsprung became engaged to Kapteyn's youngest daughter Henriëtte (Hetty), after which he left on a trip to a number of American Observatories with Kapteyn. Back in The Netherlands they made a round trip, during which they had lunch with De Sitter in Leiden, not knowing then that he would once be director of the Observatory there. They married in 1913, after which Hetty moved to Potsdam. Also in 1913 Hertzsprung published an article, in which he determined the distance to the Small Magellanic Cloud, a dwarf galaxy orbiting the Milky Way galaxy at a distance of 200,000 light-years. It was the first distance determination of an extragalactic object. He used the relation between the period and the relative luminosity of stars that are periodically varying in luminosity, a relation discovered by American astronomer Henrietta Swan Leavitt (1868–1921) in 1908. Miss Leavitt worked at the Harvard College Observatory, where Edward C. Pickering (1846–1919) was director. For her investigations Miss Leavitt had used the variable stars in the Small Magellanic Cloud, which stood all roughly at the same distance (Johnson 2006). If only the real distance of one of these so-called δ-Cepheid stars inside the Milky Way was known, the distance to the Small Magellanic Cloud could be determined. Hertzsprung achieved this making use of an extended parallax method, not taking the diameter of the orbit of the earth round the sun as base of the parallax triangle, but the much larger distance the sun traversed through space during a couple of years. He deduced from his results that a period of 6.6 days belonged to an absolute magnitude of −7.3. Variable stars in the Small Magellanic Cloud with the same period had an apparent magnitude of +13.0. From these data Hertzsprung could deduce the distance to the Small Magellanic Cloud: 30,000 light-years. This was far smaller than the real distance, but definitely outside the Milky Way. With this result we can consider Hertzsprung as a predecessor of Harlow Shapley (1885–1972), who measured the diameter of the Milky Way and Edwin Hubble (1889–1953), who determined the distances to other galaxies.

Through Kapteyn there were contacts between Hertzsprung and De Sitter. Hertzsprung wrote to De Sitter in 1915, that his photographic method could perhaps yield results with the satellites of Jupiter. He offered De Sitter to make observations for him [AH, C046/10, 21-2-1915].

After the sudden death of Schwarzschild in 1916, who had served two years in the army and had come home ill from the front, Hertzsprung asked himself where his future would be. His position as a Dane in Germany became continually more unpleasant. He missed an appointment at the small Observatory in Aarhus, because the director of the Observatory in Copenhagen, Svante E. Strömgren (1870–1947), pushed forward someone else. He had an option to be appointed at a new observatory to be built in Copenhagen. But then another possibility occurred, when his father-in-law Kapteyn offered him the possibility to become deputy director at the new astrophysical department of Leiden Observatory, after the appointment of De Sitter as director. Hertzsprung jumped at the proposal, but his patience was seriously put to the test by the slow decision process of the Dutch parliament. Strömgren wished this time to have Hertzsprung for the new observatory, but Hertzsprung also wished an astrophysical institute and an observatory on the Southern Hemisphere. But that Strömgren could not promise him. In Leiden he would get his astrophysical laboratory, and De Sitter had promised him his cooperation for the Southern Hemisphere. At the end of 1918 and the beginning of 1919 De Sitter kept Hertzsprung informed of the developments and he asked him after his plans: votes in parliament, salary, instruments to be purchased, measuring programmes. Sometimes Hertzsprung became despondent.

All that waiting, I will not speak about it anymore, until I have the appointment actually in my hands [AH, C046/10, 14-5-1919]. In May De Sitter wrote about the next delay by the Minister. After the rejection of Pannekoek he wished to wait with the appointment of Hertzsprung until a replacement of Pannekoek was known. De Sitter went mad of it all. Kapteyn had mentioned to him the Swiss astronomer L. Courvoisier, who worked in Berlin, as a possible replacement. De Sitter wished to know, if Hertzsprung and Hetty knew him, what kind of a man he was, married, children? Would they get on with him as colleague and neighbour? These personal things must not be totally ignored, although the scientific importance had of course to be preponderant. We see that De Sitter kept an eye on the human side of his possible future colleagues in his Observatory. After all the deputy directors would live more or less on top of each other. They would each even live in a part of the house directly east of the Observatory. In a postscript he ended even roguishly.

> PS, I have here a large paper cylinder for you from the Danish Academy. De post administration refuses to send it on, if they don't see what is inside. May I open it, or shall I keep it here until you come? Perhaps there is a bomb inside! But rather I presume it is a membership certificate, with which I congratulate you heartily! Tt, W. de S.
>
> [AH, C046/10, 15-5-1919; Latin tt means *totus tuus*, all yours.]

Later he wrote that he had decided not to engage Courvoisier, in order not to waste Pannekoek's last chance. He still pleaded for him with the Governors. But all his urging for haste with the Governors and the Minister was sometimes irritating.

> I ask permanently for speed, to the left, to the right, up and down, and in The Hague I think I have the name of an extraordinarily impatient bloke.

[AH, C046/10, 5-6-1919]

A month later De Sitter wrote again:

The nagging from The Hague is really unbearable.

The government in The Hague took four weeks to answer a letter and De Sitter took only three days. The Minister wished to pay only 4500 Dutch florins, while De Sitter thought that 5000 had been agreed on. In desperation De Sitter had accepted it after consulting Kapteyn. De Sitter had asked 12,000 florins for 1920 for the new instrument that would be built for Hertzsprung: the *Schraffierkassette*. It would be constructed by Carl Zeiss, a factory for optical instruments in the German town of Jena. With the instrument stars could be pictured on the photographs as a small spot, so that their luminosity could be better determined (Baneke 2005). At the time of this correspondence Pannekoek was definitively no candidate anymore. As a postscript De Sitter wrote that Hertzsprung's landlord was named Van Hoeken and that his address would be 6 Oegstgeesterlaan [AH, C046/10, 5-7-1919]. Besides about his slow appointment procedure Hertzsprung was also regularly worried about his salary, his pension rights and the fact that he had suddenly become a German by only coming to Göttingen. He wished to become a Danish citizen again and asked De Sitter if he would not suddenly become a Dutch citizen [ASL, 19-7-1919]. When Hertzsprung was in Leiden at last, it appeared that he had lost his German nationality, which made him joke:

Perhaps I can commit some evil deed, for which I cannot be punished, but it is difficult to think of something.

[ASL, 3-5-1920]

On 24 July Hertzsprung received at last the telegram he had long been waiting for:

you are appointed = De sitter

[AH, C046/10, 24-7-1919]

At last the Hertzsprungs moved, with their daughter, to Leiden in September 1919. Daughter Rigel had been born in 1916 and was named after the brightest star in the constellation of stars Orion. They still had to wait a few years before their home in the Observatory was ready. De Sitter left a month later for his long cure to Arosa in Switzerland. Suddenly Hertzsprung stood alone for his task (Herrmann 1994; Oosterhoff 1968).

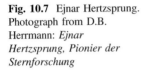

Fig. 10.7 Ejnar Hertzsprung.
Photograph from D.B.
Herrmann: *Ejnar
Hertzsprung, Pionier der
Sternforschung*

Sanatorium in Arosa and Correspondence
with Hertzsprung

One of the first tasks of Hertzsprung as acting director of the Leiden Observatory
was, at the request of Eddington, to send on the news to Einstein about the successful expeditions to Principe and Sobral. Eddington wanted to attend the colloquium in Leiden on 22 Oktober, but was unable to be there. He wrote to
Hertzsprung, who showed the letter from Eddington to Einstein, who was overjoyed with it. At the end of Oktober Einstein, Ehrenfest and Hertzsprung visited
Julius in Utrecht. Ehrenfest, Einstein and Julius' daughters made music.
Hertzsprung enjoyed himself [ASL, 27-10-1919]. Shortly after the definitive
information from Dyson arrived, from which Hertzsprung sent a copy to De Sitter.
Eddington nominated Einstein for the Gold Medal of the Royal Astronomical
Society, but he was incomprehensibly rejected. The aversion to everything German

just after the First World War was certainly one of the reasons. It was the reason that no Gold Medal was awarded in 1920. Einstein received it six years later, in 1926. Herzsprung received it in 1929 and De Sitter in 1931.

During De Sitter's stay in Arosa he and Hertzsprung corresponded extensively about all matters concerning the Observatory. Sometimes De Sitter left everything to Hertzsprung, but not seldom, as responsible director, he ordered Hertzsprung to steer a certain course. In the beginning Hertzsprung asked De Sitter's advice about nearly everything. But later he wished to leave his mark on the research policy. Both men sometimes struggled with it.

Hertzsprung started his work methodically and tried to steer all matters in detail. Soon he had his hands full with staff matters. He tried to form a picture of the younger students. Luyten[11] seemed promising, although a bit overconfident; Van den Bos,[12] who once had lost interest for a while, had to be picked from the street by instrument maker H. Zunderman. Thereafter he worked diligently but slowly at the microphotometer. Hertzsprung advised him to start working on double stars. It would be the subject of his dissertation in 1925. Smaller things kept him also busy. A new idea was that the cleaning lady could bring round a cup of coffee at 11 o'clock. Hertzsprung kept silent and the idea blew over. De Sitter was not against it, if it would make life for the staff members more agreeable, but it must not become a quarter of an hour chatting [ASL, 8-11-1919]. Was it allowed to smoke in the computer chamber? Were the computers Mien Kasten and Lena de Nie now and again allowed to be with Luyten in an observatory dome at night? It looked all innocent, but De Sitter, who had heard it from his wife, was of the opinion that Hertzsprung had to forbid it. He advised Hertzsprung in these matters to seek more often the advice from Mrs De Sitter [ASL, 6-2-1920]. Mrs Non de Sitter kept a keen eye on the Observatory and no doubt kept her husband Willem informed by letter. Sometimes she visited him. On 1 August also their children visited their father in Arosa [ASL, 2-8-1920 to Hertzsprung]. His daughter Agnes later told that they were not allowed to kiss their father due to danger of infection,[13] which points at a contagious disease like tuberculosis.

Hertzsprung had of course to become a member of the Dutch Astronomers Society founded in 1918. But he did have to pass the ballot. If I fail, it is an interesting case, Hertzsprung wrote cheerfully [ASL, 31-3-1920]. He was present at the meeting of the Society on 5 April, where Kapteyn gave a lecture about the form of the universe [ASL, 5-4-1920]. Already after a month Hertzsprung wrote he

---

[11]Jacob Luyten (1899–1994) received his doctor's degree with advisor Hertzsprung when he was 22. He worked most of his career at the University of Minnesota in the USA.

[12]Willem H. van den Bos (1896–1974) worked in Leiden and went in 1925 to the Union Observatory in South Africa, where he became director in 1941. He worked extensively on double stars. See also Oort, J., 'Obituary of Willem Hendrik van den Bos', *Proceedings of the NAW*, 1974.

[13]Private communication from Hens de Sitter, wife of De Sitter's grandson Ulbo de Sitter, Heusden (province of North-Brabant), 2008.

realized what it meant to be a director, to weigh up all kinds of petty interests [ASL, 21-11-1919]. De Sitter and Hertzsprung corresponded a lot about the new instrument that would be bought for the astrophysical department, the *Schraffierkassette*. The Zeiss company had made an expensive offer of 25,000 Dutch guilders, for the *Schraffierkassette*, the accompanying telescope and the dome [ASL, 9-12 from Hertzsprung, 1919]. Another possibility was to mount two instruments on a single support, besides the telescope for the *Schraffierkassette* also a reflecting telescope. But that would require too big a dome construction on the roof of the Observatory. It became a lengthy discussion. Another offer was asked from the Bamberg company, but they also made a bad offer [ASL, 26-4-1920]. Hertzsprung also asked the price of a dome from the English Cooke company, but they only made domes of a fixed type. Hertzsprung complained:

A German sales representative asks: *Wie wollen Sie es haben?*

An English sales representative says: *I have not got that.*

In this connection he told to De Sitter the comical story of the German who had measured that in British East India the eggs were on average one millimetre smaller in diameter than those in England, and that the egg-cups therefore could be made half a penny cheaper. The result was that the Germans had driven away the English from the egg-cups market [ASL, 12-5-1920]. Hertzsprung got the idea to appoint Kapteyn after his retirement (Kapteyn would retire in July 1921) in the Pannekoek vacancy. To be precise, Hertzsprung wrote, he had the idea on 31 March, 19:03 h in the afternoon. What do you think about it [ASL, 1-4-1920]? For the time being De Sitter considered it a strange idea [ASL, 9-4-1920.], but if Kapteyn would find it attractive, he was prepared to think about it [ASL, 23-4-1920]. Immediately Hertzsprung discussed his idea with his father-in-law [ASL, 13-4-1920], who reacted positively and wrote his ideas down [ASL, 27-4 from Hertzsprung, 1920], which Hertzsprung sent to De Sitter [ASL, 8-5-1920]. Kapteyn wrote to De Sitter he was prepared to be temporarily acting deputy director. He philosophized a bit about future candidates and even named Jan Oort already, who was only 20 years of age and advanced in his study, and Jan Schilt,[14] who was then writing his thesis in Groningen. De Sitter had planned to fetch Schilt to Leiden as an assistant [ASL, 26-6-1920 from Kapteyn]. Kapteyn's appointment would start on 1 November 1920 [ASL, 23-11-1920 from Kapteyn]. Kapteyn had become a retired member of the Academy and he proposed to De Sitter to take action in order to get Hertzsprung appointed [ASL, 8-3-1920].

De Sitter also occupied himself with the observational work. The observations with the meridian telescope, over which the observer Zwiers swayed the sceptre in the old tradition, had to be made in a different way. Not everything had to be done

---

[14]Jan Schilt (1894–1982) received his doctor's title as a student of Van Rhijn in Groningen in 1924 and worked as an assistant in Leiden from 1922 until 1925. In that year he went to the United States. He invented the later called Schilt photometer, to measure star luminosities and from that their distances. Later he worked at Columbia University until his retirement in 1962.

nigglingly accurate. A fundamental star[15] had to lose its godly status and had to be measured in the same way as ordinary programme stars. The reduction work had to keep pace with the observations. And if that was not possible, then *in God's name* less observations had to be made. Zwiers will get goose flesh from it, De Sitter wrote [ASL, 6-2-1920]. De Sitter planned to write annual reports again from 1 September 1920. He hoped there was something to communicate. Earlier reports only mentioned that part of the building had been painted, De Sitter wrote sarcastically [ASL, 22-1-1920]. He stimulated Hertzsprung to develop new activities. The outside world had to notice that after the reorganization there was new life in the Observatory [ASL, 7-5-1920], and that Leiden had woken up from its 20 years' sleep [ASL, 20-5-1920]. A completely different development was Voûte's plan to found an observatory in the Dutch East Indies. A 7-inch telescope, which was not necessary anymore in Leiden could be lent to the new observatory. The rich tea-grower Bosscha was prepared to provide money and so the plan gained momentum. At the end of May a telegram was sent to Bosscha to invite him for consultation. The question was how to make the influence of Dutch astronomy, in particular that of Leiden, on the measuring programmes as large as possible. De Sitter was emphatically active in the process of founding the Observatory in Lembang, and he tried to maximize the influence of Leiden on it. One of his main reasons was, that he had promised Hertzsprung to do his utmost to provide him with an observatory in the south [ASL, 28-5-1920 to Hertzsprung].

In May 1920 Hertzsprung thought he could infer by the spirited letters of De Sitter, that he was well again [ASL, 18-5-1920]. It appeared on the contrary that De Sitter had to stay longer and that it would take until the spring of 1921 for him to return [ASL, 23-6-1920 from Hertzsprung]. Now and again De Sitter's illness was a subject in the correspondence, in particular when there were sometimes mutual frictions. The illegible handwriting of De Sitter regularly played a role. Sometimes De Sitter could hardly hold his pen from rheumatism [ASL, 8-11-1919]. Moreover, writing was difficult in a lying position. But he would do his best. He would use a typewriter, if he had one [ASL, 16-12-1920]. Hertzsprung often wrote the transcription in his own perfect handwriting beside the illegible words. De Sitter did not keep copies of his letters. That was probably the reason that he did not always know exactly what he had asked or remarked. But he would take back Hertzsprung's letters, so that they could be united in an important archive and be published on their 200th birthday [ASL, 23-7-1920]. In March 1921 it came again to some discord between the gentlemen. On 3 March Hertzsprung wrote:

> If you write to me that I must read your letters, I can answer that you must remember (I do not underline it, because I consider that superfluous) what you have written.
>
> [ASL, 3-3-1921]

---

[15]A fundamental star was a star of which the position was measured as accurately as possible. Measurements of other stars could be related to the location of a fundamental star.

De Sitter often underlined words, once or even twice. Concerning his illness he wrote:

I am not ill for fun. ... There is no fighting against fate.

[ASL, 2-8-1920]

In May he wrote to Hertzsprung about an attempt by the physicians to apply a pneumothorax, in order to immobilize one of the lungs by inserting nitrogen in between the pleura. Probably supposing that the process of healing would be quicker. After two hours the physicians had to stop their attempts. De Sitter thought that perhaps this failed attempt would cause him to have to stay longer [ASL, 28-5-1920]. This procedure would indeed point to the contagious tuberculosis.

Hertzsprung made desperate attempts to do some astronomy, but he could not find the relaxation for it. In his opinion the money in the building process was not spent well. For the director's house of De Sitter the excellent glass artist Asperslagh was hired to design the signs of the zodiac in the roof above the central hall of the house, while everywhere else had to be economized [ASL, 10-7-1920]. De Sitter considered the stained-glass pictures also useless, but he did hope that they would not become ugly [ASL, 13-7-1920].

In the meantime, after his swearing in, Hertzsprung had become a participant of the faculty meetings [ASL, 10-7-1920]. He thought it strange that he had to swear allegiance to the queen, not having the Dutch nationality [ASL, 26-6-1920]. In July 1920 a slightly bitter tone appeared in the letters from Hertzsprung. When De Sitter went to Arosa, he had said:

Anyhow, now you are the director. I have nothing to do with it.

He had said also that he was an interested outsider. But when Hertzsprung wished to model the observing programme with the meridian circle to his own ideas, De Sitter rebelled. Hertzsprung was of the opinion that he was allowed to let others work according to his ideas, in view of the fact that he had to do all the rotten jobs and did not get around to any work of his own [ASL, 17-7-1920]. Besides the obligations he wished to have the rights of the director too. A bit later De Sitter agreed to the new meridian programme for double stars. His objection was not to have been consulted. And it committed him for a couple of years [ASL, 2-8-1920]. They had imagined everything so differently in 1919. Now Pannekoek wasn't there, De Sitter was in Switzerland, Hertzsprung lived far from the Observatory and had not yet money for his new telescope. He did not want to make reproaches, but he did wish that his position was taken into account [ASL, 28-7-1920]. De Sitter's opinion was that he had to lay down the outlines himself. He wished to really be the director [ASL, 2-8-1920]. He had preferred that someone else had become director, but now he had to take his responsibility [ASL, 13-7-1920]. He even wrote that he would have had no objection against Hertzsprung having become director, and he deputy director. But that was not the case, and responsibility implied authority.

I would prefer not to interfere, but I am not permitted to do that.

[ASL, 23-7-1920]

In the summer Hertzsprung went for a well-deserved holiday to Denmark and turned over the direction to Zwiers [ASL, 26-6-1920]. A busy six months followed after the holidays until De Sitter's return. In the meetings of the Dutch Astronomers Society the building of the observatory in the East Indies, in Lembang, near Bandung on Java, was often discussed. The gist of the discussions was that it had to be prevented that Leiden would be the only in charge. Kapteyn supported this point of view, against De Sitter's wishes [ASL, 4-1-1921 from Hertzsprung]. In one of the meetings Van der Bilt presented the idea to build one large Dutch observatory on the Veluwe, in the province of Gelderland. The Veluwe is a nature reserve and was the perfect location for an observatory, in those days one of the darkest areas during the night in The Netherlands. Nobody paid much attention to the idea and there was hardly any discussion about it [ASL, 29-9-1920]. Perhaps it was a last sign of Van der Bilt's irritation not having been asked for a position in Leiden. In the meantime the 7-inch telescope was on its way to the Bosscha Observatory in Lembang in the East Indies. Hertzsprung found that a number of test measurements had to be made first to ascertain if the air was such that parallax measurements could be made. Negotiations had been started about the construction of a large telescope for the Bosscha Observatory [ASL, 22-10-1920]. At the end of 1920 De Sitter wrote that he wished to take the lead in the Lembang case. And if that was not possible, he would completely distance himself from it. Not from vanity or ambition, but because he had to spend all his time to Leiden [ASL, 16-12-1920]. Of course Hertzsprung was not satisfied with De Sitter's threat to withdraw from the Bosscha Observatory, because it was one of De Sitter's promises to create measurement possibilities in the south. Later he was less adamant and changed his rather blunt remark into having an important say in the matter [ASL, 6-1-1921].

At last the *Schraffierkassette* was ordered in October 1920. Hertzsprung must have breathed a sigh of relief, when this condition to come to Leiden was fulfilled. In view of the coming retirement of Kapteyn De Sitter had conceived the idea to honour him internationally with a publication of his collected papers [ASL, 10-12-1920 from Hertzsprung]. But he had to abandon this idea for lack of interest. Hertzsprung sent a few photographs of the building activities on the director's house in November, which made fast progress. Hertzsprung proposed the idea of starting an astronomical magazine. The German *Astronomische Nachrichten* had lost part of its significance and the English wanted to stay exclusive [ASL, 15-2-1921].

In early April De Sitter returned to Leiden. He himself presented the budget of the Observatory for the year 1922 to the Board of Governors a few days later. Hertzsprung had asked for a postponement [ASL, 14-3-1921 to Hertzsprung]. Pannekoek, very happy with his position in Amsterdam, welcomed him back and wrote him not to work too hard yet [ASL, 6-4-1921]. When De Sitter was in Arosa, Pannekoek had written him a few times to ask for explanation about the general theory of relativity, which De Sitter had given him [ASL, 10-11-1919, 24-1-1920]. Everybody was glad about De Sitter's return, also Elise Kapteyn, who wrote to him:

Dear Willem, Will you please be very careful now, and not walk through draughty corridors?

[ASL, 22-4-1921]

It may be true that a few times mutual irritation had arisen between De Sitter and Hertzsprung, but the general tone of the correspondence during this special year and a half was always one of mutual respect and the wish to inform the other as honestly and completely as possible of the events, ideas and opinions. It must have been very hard for De Sitter to execute the duty he felt as the responsible director, while being an ill man, writing from a large distance and having hardly any experience as a director. Hertzsprung's devotion to duty and honesty count highly in his favour, because he not seldom wrote several times a week to inform De Sitter of the ins and outs of the Observatory. The result was that De Sitter could actually smoothly take *power* again in April 1921.

The above survey may elucidate that De Sitter had not at all lost feeling with nor influence on his Observatory. For all this he owed a lot to Hertzsprung, something that can hardly be overestimated.

References

Baneke, D.M. (2005). 'Als bij toverslag. De reorganisatie en nieuwe bloei van de Leidse Sterrewacht, 1918–1924'; *Bijdragen en Mededelingen Betreffende de Geschiedenis der Nederlanden*, 120 (2), pp. 207–225, 2005. ('Like magic. The reorganization and new bloom of Leiden Observatory, 1918–1924'; *Contributions and Communications Concerning the History of The Netherlands.*)

De Sitter, W.R. (2000). 'Kapteyn and De Sitter; a Rare and Special Teacher-Student and Coach-Player Relationship'; in: P.C. van der Kruit and K. van Berkel, editors, *The Legacy of J. C. Kapteyn, Studies on Kapteyn and the Development of Modern Astronomy*, Kluwer Academic Publishers, Dordrecht Boston London, 2000. (*Astrophysics and Space Science Library, Volume 246.*)

Herrmann, D.B. (1994). *Ejnar Hertzsprung, Pionier der Sternforschung*, Springer-Verlag, Berlin Heidelberg, 1994.

Johnson, G. (2006). *Miss Leavitt's stars*, Norton & Company, New York, 2006.

Oosterhoff, P.Th. (1968). 'Levensbericht ('Life Report') E. Hertzsprung'; *KNAW, Jaarboek (Annual) 1967–1968*, pp. 322–325.

Pannekoek, A. et al. (1982). *Reminiscences*. See References of chapter 1.

Chapter 11
At the Top

Also I remember your frequent statement that one principal function of your Observatory is to train and export highly capable young astronomers.

Harlow Shapley to Willem de Sitter in 1928
[ASL, 22-12-1928]

Decade of Organization and Management

After the decade of his main theoretical achievements De Sitter's appointment as director of the Leiden Observatory had caused a great career change. After his return from Arosa the emphasis of his work shifted to organization, administration and management. First there was the reorganization of the Observatory. After its successful completion De Sitter became secretary of the Senate of the University and the following year he was Rector of the University.

He was chosen as vice-president of the International Astronomical Union at its general assembly in Rome in 1922. His election as president followed in 1925, which implied also many international contacts. It culminated in the organization of the general assembly in Leiden in 1928. Two other international projects took a lot of De Sitter's time too. First the execution of the partnership agreement, reached with the Union Observatory in Johannesburg in South Africa in 1923. Second the equator expedition to Kenia in Africa in order to measure accurate declinations during 1931–1933. All these activities explain that De Sitter was less active in those years in the field of pure science, in particular cosmology. This elicited the remark from Eddington that De Sitter had invented a universe, but subsequently had forgotten it (Eddington 1934). It was not until 1929 before he picked up where he left

© Springer Nature Switzerland AG 2018
J. Guichelaar, *Willem de Sitter*, Springer Biographies,
https://doi.org/10.1007/978-3-319-98337-0_11

Fig. 11.1 De Sitter, working and smoking at his desk. Notice his unusual pen grip

off in 1917. In this chapter the reorganization of the Observatory and his work in the Senate will be discussed and the international activities in the next chapter.

Reorganization of the Observatory[1]

At the request of the Board of Governors De Sitter had already, after consulting Kapteyn, put to paper his ideas for the reorganization. A *cheerful scientific life* with measuring programmes of high quality reigned up to 1900 at the Observatory, although many observations had not yet been published. After 1900 also much less observations were made. According to De Sitter the situation had to be called very unsound. There were too less computers in Leiden to reduce the measurements. De Sitter was extremely critical:

> The new photographic refractor, which was placed in 1897, has not yet yielded any result.

In the annual reports of the Observatory technical improvements were mentioned every year, so that, De Sitter wrote not without sarcasm:

---

[1]See also Reference (Baneke 2005) in Chap. 10.

The reader gets the impression in the end, that this must be the most perfect instrument in the world.

But observations were absent. The set of instruments and the domes were out-of-date. The clock, the Hohwü 17, had not been cleaned since 1898. De Sitter had already ordered a new cleaning. There had to be made room for a library and archive, computers and the measuring of photographic plates. Moreover, there had to be installed a connection to the electricity grid. Three deputy directors had to be appointed for the three departments, with good salaries and accommodation. Not the student dorms, where the observers then lived. There had to be created a number of assistantships, and it should be possible that computers also became assistant-observers. The budget for equipment had to be raised drastically. Moreover, there had to be extra money for a process of catching up. So far De Sitter's report [AUL, BG, 17-4-1918].

The first renovation programme was made by the government architect according to the instructions of De Sitter. The estimated investment was 300,000 Dutch guilders. The costs at the end of the process would be significantly higher. On top of the house on the west side, for the director, and the one at the east side, for two deputy directors, a storey would be built. And for the observers and other staff members a few new houses would be built on the strip of ground along the entry road to the Observatory [ASL, 13-1-1919 from the architect]. During De Sitter's stay in Arosa a start had been made with the building. Not everything went well. Observer C.H. Hins (1890–1951) wrote that the previous archives room had collapsed and that the chipping had caused so much dust that Zwiers had had the meridian circle sealed up [ASL, 29-7-1920].

At his return De Sitter was dissatisfied with the progress. Already a few days after his return he wrote a strongly worded letter to the government architect. The progress was too slow. There were only a few labourers at work. Elsewhere there was more money to be earned, so the wages had to be raised substantially. The west wing with the director's home had to be finished in the summer [ASL, 9-4-1921]. Like the appointment procedure it became an agony. In 1922, again a year later, permanent secretary Feith, who had opposed De Sitter in 1918, asked if the central heating could not be abandoned [ASL, 26-4-1922]. De Sitter, then in Rome for the meeting of the International Astronomical Union, answered that, due to the very disturbing smoke, a central heating was absolutely necessary [ASL, 4-5-1922]. The teasing went even further. Could there not be installed a regular clock installation? And was an internal telephone system really necessary? Again De Sitter wrote, also to the government architect, that everything was necessary. The time installation was necessary for the observers. And due to the size of the buildings the telephone system was also necessary. Perhaps a few apparatuses could be missed [ASL, several letters, 1922]. De Sitter resisted also the plan to execute the whole building process in three years. It had to be done in two years. In the end it was 1924 when the rebuilt and reorganized Observatory could be opened officially.

Departments

The new structure with three departments was introduced: the astrophysical
department under the management of Hertzsprung, the theoretical department under
De Sitter, in which his student Woltjer played an important role, and the fundamental or astrometric department, for the time being led by Zwiers, originally
Pannekoek's function. According to Hertzsprung's idea Kapteyn became deputy
director for one day a week. De Sitter had decided already in 1919 to stop the
programme of measuring fundamental stars with the meridian circle and to replace
it by a programme of position measurements of a number of red stars. He charged
Hins with this task. Hins, although interested at first in a theoretical subject, was
willing to write his dissertation based on his observations. De Sitter agreed to this
from Arosa [ASL, 12-2, 24-4 from Hins, 1920].

Hendrik J. Zwiers (1865–1923)
Zwiers started working at the Observatory in 1891 and received his doctor's
degree in 1895 with a thesis on the comet Holmes. He developed a method to
calculate the orbits of double stars. In 1907 he became an observer at the
Observatory, in which function he was responsible for the meridian observations for many years. He was also in charge of the computers of the
Observatory. Outside his scientific work he was active in the city council of
Leiden for thirteen years, as member of a liberal party. He died after a short
sickbed in 1923. (Leidsch Dagblad, 13-12-1923)

Coert Hendrik Hins (1890–1951)
Hins started his studies in Leiden in 1908. He became an observer at the
Observatory. He received his doctor's degree in 1925, with De Sitter as
advisor. The title of his thesis was *Introduction to a catalogue of positions and
proper motions of 1533 red stars* (Hins 1925). The work was done as part of
the first meridian programme under De Sitter's directorate. Position measurements of these 1533 red stars, together with all older observations, had to
yield the proper motions of these stars. Hins was one of the last astronomers
who applied himself to the traditional position measurements of stars and the
accompanying techniques. He also measured 1073 stars which were used by
Schlesinger as fundamental stars. Besides he made a catalogue of 1172 reference stars for Kapteyn's Selected Areas programme. During the general
assembly of the International Astronomical Union in Leiden in 1928 Hins was
local secretary. Hins was leader of the Kenia expedition in the years 1931–
1933, when he measured fundamental declinations of stars in cooperation with
assistant G. van Herk (1907–1999). The instrument they used measured corrections for the systematic inaccuracies in the declination measurements up to
that time. Hins had been in the military service during the First World War. In

the Second World War he took part in the resistance movement. After the war he went to Java in 1946 to help rebuild the damaged Bosscha Observatory. There he was appointed professor in 1947. He received a royal award. Hins was a flamboyant man and a fanatic chess and bridge player. He died in Leiden in 1951, having been run over by a bus.[2] (Oort 1951)

A catch-up effort was made for reduction and publication of old observations. One of the publications was the fundamental position measurements of 84 pole stars from the years 1877–1885, printed in 1922 on the occasion of the 50 years' jubilee of the acceptance of the directorate of the Observatory by H. G. van de Sande Bakhuyzen. After the death of Kapteyn Leiden took over part of the work on the Selected Areas. The first successes did not stay unnoticed internationally. De Sitter wrote to the president of the Board of Governors De Gijselaar, taking a rest in Florence after his visit to the general assembly of the International Astronomical Union in Rome, that not only the assembly had been a great success, but that the Dutch delegation had been received with a lot of recognition. The Netherlands were considered on an equal basis among the great astronomical countries, and in particular the Leiden Observatory was praised for its reorganization. A Dutchman[3] was appointed vice-president of the IAU. In order to comply with all international arrangements about measuring programmes, it was thus necessary to fully implement the reorganization. De Sitter asked De Gijselaar again to talk again with the Minister in order to put all that was promised on the budget for 1923. De Sitter himself had to recover of all the fatigues and asked for an extra two weeks' leave [ASL, 14-5-1922].

Inauguration of the Reorganized Observatory

At last the day had come on 18 September 1924 that the reorganized Observatory could be inaugurated. It was done by the Minister of Education De Visser (De Visser 1924). He had given notice beforehand that the programme had to be finished at 11 o'clock. Still late, because the Minister usually went to bed at 11. After the speeches there would be a tour through the Observatory with, when the sky was clear, a look through one of the telescopes. De Sitter had to send in a typed version of his speech in advance. The Minister, member of the political party Christian Historical Union[4], spoke first and opened the reorganized Observatory with the words *God only is great* (De Visser 1924). De Sitter himself came after a short

---

[2]'Prof. dr. C. H. Hins†', article in the paper *Leidsch Dagblad*; 22 October 1951.
[3]De Sitter himself.
[4]Dutch: Christelijk-Historische Unie.

introduction about the merits of Kaiser and the old Van de Sande Bakhuyzen to the core of the reorganization.[5] First he had decided to stop with the fundamental position measurements. They took a lot of time; moreover, the modern large telescopes were better equipped for that purpose. Differential position measurements in the fundamental department with the meridian telescope, relative to the exactly determined fundamental stars, were at least as valuable and took much less observation and reduction time. A first series (the 1500 stars measured by Hins) had already been completed. Further De Sitter had implemented a modification in the organization, creating the possibility that computers, after an additional training, could also perform part of the observations. De Sitter called this a transition from the German to the English system. Partly due to this transition Hins' programma had been completed so fast. In a next subject De Sitter asked if the time had not come to look for other methods and instruments, besides the meridian circle. He hinted to the old ideas of Kapteyn, that azimuth measurements at the equator could lead to more accurate declinations. Hins and Van Herk would leave for Kenia in 1931 with a new azimuth instrument for these measurements. For the new astrophysical department of Hertzsprung a new instrument was the double camera with a movable plate holder, built by the Zeiss company and recently placed. The 13-in. photographic refractor, bought in 1897 and the largest telescope in Leiden, also belonged to the astrophysical department, although it was not specifically designed for the department programmes. Next to the domed tower for this telescope a small building had been constructed, with studies and dark rooms for the department. The field of research of the department was double stars and photographic photometry with the help of a specially constructed microphotometer. However, the equipment of Leiden was modest. For further investigations observations from observatories abroad were necessary. These measurements, mainly photographic plates, were done for Leiden in great numbers. Perhaps more important was the cooperation agreement made in 1923 with the Union Observatory in Johannesburg. According to this agreement there was a mutual right for staff members to work for a prolonged time with the colleagues on the other continent. In the meantime Hertzsprung had been already in Johannesburg for a couple months as first Leiden astronomer. In this way Leiden did not only have a much larger telescope at its disposal, but they also had the best observing conditions. The third department was the theoretical department, led by De Sitter himself. In fact, this department had existed since the appointment of De Sitter in 1908, being outside the Observatory organization. But now it was extended and part of the Observatory. The publications had also been increased enormously, certainly helped by the start a few years earlier of the *Bulletin of the Astronomical Institutes of The Netherlands*, which had a large international circulation. For large publications the *Annals of the Observatory in Leiden* were used.

---

[5]Address of De Sitter. Present in ASL.

In conclusion we can state that in 1924 Leiden was again an internationally respected observatory with sound and up-to-date research and a staff of high quality.

Staff and Employees of Leiden Observatory in 1924
Prof. Dr. W. de Sitter, director, theoretical department.
Prof. Dr. E. Hertzsprung, deputy director, astrophysical department.
Dr. J. Woltjer, assistant, theoretical department.
J. H. Oort, assistant, fundamental department.
C. H. Hins, observer, fundamental department.
W. H. van den Bos, observer, astrophysical department.
Dr. J. Schilt, assistant, astrophysical department.
D. Brouwer, assistant.
Mrs. J. M. Bruggeman, assistant.

Further there were eight computers (and another five, who worked without an appointment), two instrument makers and one stoker/carpenter. Of the staff of six under Van de Sande Bakhuyzen only Woltjer was left.

Secretary and Rector of the Senate of Leiden University

De Sitter was appointed by the Senate of Leiden University as its Secretary on 15 September 1924. His successful activities to reorganize the Observatory had doubtlessly not escaped the attention of the colleagues. The Senate was the daily scientific administration of the university. The Senate consisted of all professors, but was subordinate to the Board of Governors. The University was granted in 1575 by the Prince of Orange to Leiden from gratitude for the long resistance against the Spaniards ending in the so-called Leiden's relief.[6] The appointment of the Rector was reserved in the first period of the University to the Orange family. All professors were considered to be present at the meetings of the Senate. For a number of centuries a professor who did not appear was fined. Before being appointed for the period of one year it was customary to be Secretary of the Senate in the preceding year. De Sitter spent a lot of time on his work for the Senate. On most working days it took him nearly the whole afternoon. The morning was devoted to managing the Observatory, so that for astronomical research only the evening remained. In the following academic year 1925/1926 De Sitter was Rector of the University. At the end of September he received his dignity as Rector by his predecessor A.J. Blok.

---

[6]Leiden was sieged by the Spaniards in the years 1573–1574 during the Eighty Years' War. The siege ended on 3 October 1574. Each year the relief is celebrated on 3 October with the eating of herring and white bread.

Fig. 11.2 Honorary doctorate of Queen Wilhelmina in 1925. Secretary of the Senate Willem de Sitter is seated on the far right. To his right are seated Princess Juliana, Queen Mother Emma, Queen Wilhelmina and Prince Consort Hendrik

That implied a number of representative obligations. The great enthusiasm with which De Sitter threw himself into this representation of the Senate is amazing. The many newspaper reports bear witness to an innumerable number of visits at social, scientific and student gatherings. A few highlights from these two Senate years are worth mentioning.

Wilhelmina, Queen of The Netherlands received an honorary doctorate on 9 February 1925 on the occasion of the 350th anniversary of the University. De Sitter, in his function of Secretary, with cap, gown and insignia, conducted Queen Mother Emma to her seat in the Pieter's Church. On a photograph of the scene De Sitter is seen keeping the degree certificate for Wilhelmina in his hands during the speech of Rector Blok. Wilhelmina not only received her doctorate in the science of law due to the historical ties of the House of Orange with the University, but also for her practical endeavours to implement law nationally as well as internationally. After the official formalities the royal guests were received later by the Rector and the Secretary in the Senate chamber. The Queen also spoke with Einstein. She left at ten to five, with her doctor's certificate under her arm, signed by De Sitter. The Queen offered a dinner on 26 February as a token of gratitude, to which De Sitter was also invited.

For De Sitter personally the speech which he delivered as Rector on 11 December 1925 for Lorentz, whom he admired very much, was important. The reasons were the 50 years' jubilee of Lorentz' doctorate and the reception of an honorary doctorate in the medical sciences by Lorentz. In his address De Sitter said:

> In The Netherlands the number of eminent men of science per square kilometre is larger than anywhere else in the world.

On the *dies natalis* or foundation day on 8 February 1926 Rector De Sitter delivered the oration, titled *The unity of science* (De Sitter 1926). Science had been successful in so many fields, that it seemed to have developed into chaos. However, according to De Sitter the unity was hidden in the aim and the means. The aim: the search to the reality behind the phenomena. The laws of nature, stripped of the influences of the observer, were the reality. The means: to formulate hypotheses and testing those by means of observations. De Sitter gave an exposition at the hand of important astronomical fields of research: the structure of the universe and the evolution of the stars, or the distribution of matter in space and the properties of matter. A third field of research, gravitation, had shifted a bit to the background in those days, but its time would come again. A few other remarks in the oration deserve attention. First the fact that energy and matter were one according to the theory of relativity: $E = mc^2$. That fact led to the possibility that matter (electrons and protons) could disappear and be converted into radiation energy. With a rare clear view into the future he added:

> Let us hope that humanity will not succeed to realize this, before the currently used methods to settle disputes between large groups of men will belong to a long forgotten barbaric past.

His hopes did not realize. Nineteen years later two atomic bombs would end the Second World War. According to De Sitter this source of energy could very well be the source of radiating stars. It was further necessary for the unity in the astronomy to do very many small observations, with patience and organized cooperation:

> It is the chain that is important, not the single links.

At last De Sitter said that the delight of knowledge and of the road to more insight was of a higher order than the mystic exaltation of the artist; and that mathematics was the means to free ourselves from our limitations and was the stairway to the unattainable miracle.

With a speech on the events in the University over the past year De Sitter turned over his office to the new Rector E. M. Meijers on 20 September 1926.

A New Top Observatory

The last annual report concerning the activities at the Observatory of Van de Sande Bakhuyzen dealt with the year 1912. This lagging behind was already an indication of the decline of the Observatory. De Sitter took up this task again from 1 May 1919.

The first striking fact from the annual reports up to 1933 is that the Observatory had quickly become a well-oiled machine.[7] De Sitter put a lot of time into the recruitment of new and capable staff members, and to the terms of employment of the personnel, certainly also of the lower ranks. It was clear to everybody, from high to low, who was the boss and who made the decisions. De Sitter listened to his personnel, but made the decisions alone and then acted consistently. He treated everybody with respect and sympathy. Once when an employee came nervously to De Sitter to confess a mistake, the only thing De Sitter said reassuringly was that only people who worked could make mistakes, repeating a one-liner of David Gill. De Sitter also tried to improve the image of the Observatory. Visitors from the citizenry were welcome. Once a week De Sitter opened the doors of the Observatory in the evening for visitors, who were allowed to have a look through one of the telescopes if the skies were clear. The number of visitors increased from 465 in 1923 till a maximum of 1812 in 1931, after which it decreased till 800 in 1935. The peak in 1931 was probably due to the great public interest for the expanding universe, with which De Sitter regularly reached the press. Checking the possessions of the Observatory was also necessary. The 18,000 books in the library were compared to the catalogue yearly. Sometimes some books were missing, but those appeared again most of the time at the next verification. De Sitter considered it important enough to mention it in the annual report.

De Sitter paid much attention to the maintenance and the precision adjustments of the instruments. To that aim a lot of work was done in the technical department of the Observatory, which was led by H. Zunderman. Sometimes an advanced student from the instrument makers school of Kamerlingh Onnes worked in the technical department. A problem had to be solved with the meridian circle, which had been turned slightly relative to the meridian marking signs. This had been caused during the building renovation by the installation of new concrete poles besides the meridian room. Using measurements at the Pole Star the correction was regularly determined. The meridian circle (after Hins´ measurements at the red stars), the photographic refractor and the 10-inch visual refractor were completely cleaned and overhauled in the year 1922. Servicing and overhauling of the instruments were often performed yearly, if there was time between the measuring programmes. The instrument specially made for Hertzsprung by Carl Zeiss in Jena came to Leiden not until 1924. It consisted of two identical telescopes with a focal distance of only 52 centimetres and objectives with a diameter of 104 millimetres. With the two telescopes photographs could be made in two different parts of the colour spectrum. On plates of 20 centimetres squared photographs could be made from a solid angle of $20° \times 20°$. The plateholders could be shifted by electromagnets, causing the light of a star to spread out over 1 square mm. The movements resembled engraving (German: *schraffieren*), from which the name *Schraffierkassette*. In this way the luminosity of a star could be measured much better than on the basis of a single point, which the starlight left on the plate without

---

[7]The annual reports are present in ASL.

shifting. This method had been used for the first time by Schwarzschild. The telescope was thus very suitable for the mass of research Hertzsprung did on variable stars, in which the luminosity had to be determined on many points in time to determine the luminosity curve.

Observations, Instruments, Reductions and Theory in Leiden

Observations with the instruments of the Leiden Observatory were no sinecure. A clear sight on the skies was often wholly or partially made impossible by a blanket of clouds. Rain and cold were also often playing tricks with the observers. Often all preparations for a series of observations had been made, when suddenly a number of clouds made it all worthless. Exactly for necessary observations the weather could be a formidable spoilsport: passings of the meridian of particular stars, or photographs of variable stars necessary to determine intensity curves. Working during the night must not be underestimated either. Measuring mostly at night and working as much as possible during the day, there were no ordinary working hours for the passionate astronomer.

Up to 1925 the positions of stars in a number of Selected Areas of Kapteyn were measured (areas of $2° \times 2°$ spread out over the 24 h of the right ascension and with declinations of 0, 15, 30, 45, 60 and 75 degrees). Each area contained roughly ten stars. They always were differential measurements. After that it took a number of years to reduce the observations. Under the guidance of Hertzsprung a varied programme of measurements was executed for years: with the photographic refractor the Jupiter satellites, double stars, parallaxes and variable stars, and with the 10-inch visual refractor double star measurements, mainly by Van den Bos. The assistants Luyten, Van den Bos and Schilt were very active in those first years of De Sitter's directorate. In later years Van Gent[8], Oosterhoff[9] and Kuiper[10] followed.

> **Scientific staff from 1919 until 1934**
> ** J. Woltjer, observer until 1919, curator from 1919, lecturer and later professor from 1922.

---

[8]Hendrik van Gent (1900–1947) was a student of Hertzsprung (dissertation in 1932) and went to Johannesburg in 1928. His main subjects were the study of variable stars and planetoids.

[9]Pieter Th. Oosterhoff (1904–1978) was a student of Hertzsprung (dissertation in 1933). He discovered populations of spherical star clusters, the Oosterhoff groups. He was secretary-general of the International Astronomical Union from 1952–1958.

[10]Gerard Kuiper (1905–1973) was a student of Hertzsprung and worked with him on double stars. He went to America and worked at several observatories and became an American citizen. He supposed the existence of a belt of objects of rock and ice at a distance of 30 till 50 astronomical units from the sun, later called the Kuiper Belt.

** J. Weeder, curator until 1922.
** H. Zwiers, observer, died in 1923.
** W. de Sitter, director from 1919.
** E. Hertzsprung, deputy director from 1919.
** C. Hins, observer from 1919.
** W. Luyten, assistant in 1920.
** J. Kapteyn, temporary deputy director from 1920, died in 1922.
** W. van den Bos, assistant and observer from 1921 until 1925.
** J. Schilt, assistant from 1922 until 1924.
** W. Kruytbosch, voluntary assistant from 1923 until 1933.
** J. Oort, curator from 1924.
** D. Brouwer, assistant from 1924 until 1926.
** J. Bruggeman, in 1924.
** H. van Gent, assistant from 1925 until 1927, observer from 1928.
** Doorn, assistant from 1925 until 1927.
** C. Sanders, chief assistant from 1926 until 1931.
** J. Raimond (1903–1961), assistant from 1927 until 1928.
** G. Kuiper, assistant from 1928 until 1934.
** P. Oosterhoff, assistant from 1928.
** A. de Sitter, assistant from 1929.
** G. van Herk, chief assistant from 1932.
** A. Wesselink, assistant from 1932.

The moving-plate camera of Hertzsprung gave quite a number of problems during the first years and had to be readjusted regularly. But after a few years it performed excellently. The luminosity changes of variable stars were a considerable part of the observations. The luminosities were determined with the Schilt photometer. Under the guidance of instrument maker Zunderman the 't Hart company in Rotterdam constructed a number of Schilt photometers for a number of other observatories. Also under the guidance of Zunderman a start was even made with the grinding of a mirror with a diameter of 40 cm for a reflecting telescope.

Refractor and reflector
A *refractor* is a telescope based on lenses. After refraction by the objective and the ocular lenses the light rays form an image. The largest scientific refractor is the one of the Yerkes Observatory near Chicago, with an object lens diameter of 102 centimetres. Larger refractors are hardly possible to make due to the bending of the lens, because it can only be supported at the rim.

A *reflector* is a telescope with which the primary image is made by reflection by a hollow mirror. Isaac Newton was one of the first scientists to construct one. The development to always larger mirrors with diameters of many metres is still continuing. Large mirrors are possible by the use of thinner reflecting materials, smaller components and correcting supports in many points.

In order to make the observations the time signals were of the utmost importance. After the telegraphic time signals had been stopped, they were received by radio. In 1924 there came a new receiver with an antenna of 125 metres long. But the signals were sometimes hard to receive. The most prominent national receiver and transmission station in Kootwijk (province of Gelderland) had sometimes difficulty in receiving the signals. Moreover, sometimes Kootwijk had the irritating habit of transmitting telegrams on the frequencies and times of the astronomical time signals.

After the measurements on the Selected Areas the astrometric department executed a measuring programme of reference stars for Schlesinger for a couple of years.

In the theoretical department a lot of work was going on too. De Sitter himself put a lot of time in his Galilean moons. And the variations of the earth's rotation were also a subject of his theoretical studies. From 1929 on cosmology came in view again. De Sitter did research on apparent diameters and magnitudes, distances and radial velocities of extragalactic nebulae, partly based on data from Hubble. The correlation between the measured radial velocities and the distances led to the concept of the expanding universe, in agreement with the solution of the Einstein equations given by the Belgian astronomer Lemaître.

After his work on celestial mechanics and the study of the satellite Hyperion of the planet Saturn Woltjer applied himself to Eddington's star theory. Among other subjects he worked on the periods of pulsating Cepheid stars and on the chromosphere of the sun.

Oort did statistical-theoretical research, although he was part of the astrometric department. He made study of stars with high velocities, on which he wrote his dissertation. Later this study would lead to a description of the rotating Milky Way. De Sitter's son Aernout, in the meantime busy with his study of astronomy, supported Oort with parts of his enquiries. In order to determine the movement of the solar system Oort deduced that this was nearly completely dependent on the rotation of the galactic system. From the distribution of the star velocities perpendicular to the plane of the Milky Way and the variation in the star density in the same direction he tried to deduce the working forces.

The number of publications in those years was large. In contrast to the period before De Sitter's directorate observations, reductions and publications kept in pace.

A cooperation agreement was signed with the Union Observatory in Johannesburg in 1923. From that time onwards a Leiden astronomer was always working at the observatory of De Sitter's friend Innes, later of his successor the English astronomer H. E. Wood (1881–1946).[11] Hertzsprung was the first one. In the 1920s preparations were made to measure accurate declinations of a large number of stars. To this end a method was used invented by amateur astronomer C. Sanders, who had performed measurements with this method in Matuba in Portuguese Congo in Africa.

His measurements were reduced in Leiden. Sanders worked as chief assistant from 1926 until 1931 at the Observatory. During 1931–1932 Hins and Van Herk went for over a year to a temporary observation post very near the equator in Kenia, in order to do measurements of star declinations according to Sanders's method using an azimuth instrument specially devised for the purpose. In the next chapter this and other international activities of De Sitter and the Observatory are described. The Observatory knew a fifteen year period of prosperity and success during De Sitter's reign.

De Sitter's Qualities

The question arises which were De Sitter's qualities leading to the renewed success of the Leiden Observatory. First there was De Sitter's personality. His descent and upbringing in an intellectual-governmental environment provided him with a nearly innate self-assurance. His first correspondence in 1896 with Gill showed De Sitter, although still a student with hardly any astronomical experience, as a self-confident young man, who was able to write an extensive measuring programme for his period at the Cape Observatory. As a next point springs forward a total absence of even the least amount of subservience to his superiors, both in the astronomical as in the administrative field. He always wrote with respect to and about Gill and Kapteyn, but did never hesitate to give his own opinion. The same holds true for his letters and comments in his controversy with Sampson, although he had only recently obtained his doctor's degree. In his letters to the Board of Governors he stated his wishes straightforwardly and without any servility. There is a world of difference between De Sitter's letters with wishes, that hardly tolerated objections, and the somewhat lamenting style in which his predecessor Van de Sande Bakhuyzen wrote his wishes for the budget, of which hardly ever something was met. On photographs with colleagues we always see De Sitter proudly stand upright, often with a cigarette or a pipe nonchalantly in his hand. He stopped with the yearly filling out of a number of forms, a duty for all institute directors, for the Board of Governors, stating that he published an annual report about the Observatory, in which all relevant data could be found. De Sitter did not lack the conviction of his own right either. When he had taken a decision, he acted without

---

[11]The first three directors of the Union Observatory were Innes (1903–1927), H. E. Wood (1927–1941) and Van den Bos (1941–1956).

hesitation accordingly. He had hardly any doubts and took the possible adverse consequences of his decisions for granted. Sometimes it led to difficulties. During his trip to The United States at the end of 1931 he transferred part of his responsibilities to Oort and not to Hertzsprung, who, as deputy director, was the right man to take over temporarily. Why he did this, is not clear. It must have hurt Hertzsprung. But the consequences of his decisions did not leave him cold. His wife wrote that in times of tension he worried and as a result had a restless sleep.

Another quality of De Sitter was his capability to raise funds. In order to organize the expedition to Kenia in 1931–1932 to measure fundamental declinations of stars he was able to raise several fundings, amongst others from the Holland Society of Sciences (HMW) and from the International Astronomical Union. His largest success in this field was the grant in 1930 of $110,000.00 from the Rockefeller Foundation for the construction of a large telescope, which would be built on the grounds of the Union Observatory near Johannesburg.

Another reason for his success as the Observatory director was the fact that, at the time of his appointment, he could already pride himself on an important service record. His epoch-making articles in *The Observatory* and *The Monthly Notices* of the RAS had established his name as a leading theoretician, particularly amongst the Dutch and English physicists and astronomers. His theoretical work in celestial mechanics on Jupiter's satellites had contributed too. That was one of the reasons that he could, supported by the settled tycoon Kapteyn, recruit excellent astronomers. Hertzsprung, already before De Sitter a famous astronomer and married to a daughter of Kapteyn, had come from Potsdam to Leiden with pleasure. Pannekoek would also have liked to come, but was not allowed to. Amongst rising astronomers it soon became obvious: Leiden was the place to be with its two great astronomers. From Groningen Oort and Schilt came to swell the Leiden ranks. In Groningen they were not amused.

Jan Hendrik Oort (1900–1992)

Jan Oort received his master's degree as a student of Kapteyn in Groningen already in 1921. Then he worked for two years at the Yale Observatory in the United States as an assistant of Schlesinger, after which he obtained a position at the reorganized Observatory in Leiden. He defended his dissertation as a student of Van Rhijn in 1926, titled *The stars of high velocity*. The velocities of stars with a high velocity were not arbitrarily oriented, like those with low velocities. After a theoretical suggestion of the Swedish astronomer Bertil Lindblad (1895–1965) about differential rotation of the Milky Way, Oort realized that he could confirm these ideas with his observations. He published his result in 1927 and spoke about it on the Ehrenfest colloquium on 4 May [AL, 3-5-1927 Oort to Lorentz]. After De Sitter's death he became deputy director and extraordinary professor under De Sitter's successor Hertzsprung. After Hertzsprung's retirement, retarded by the Second World War, he became director of the Observatory in 1946. In the management of the Observatory he was supported by Oosterhoff. After the war Oort became the *father* of the developing radio astronomy. That all started after Van de

Hulst, on a suggestion of Oort, had calculated that the 21-cm line from the hydrogen spectrum could probably be measured. This electromagnetic wavelength in the radio spectrum lay outside the visible spectrum. Up to then no other wavelengths could be measured, because they could not traverse the earth's atmosphere. But the radio waves could reach the earth. And the 21-cm line was strong enough to be measured. From 1951 onwards the structure of the Milky Way was mapped. Oort was general secretary of the International Astronomical Union from 1935 until 1948 and became a member of the Dutch Royal Academy of Sciences in 1937. After his retirement he stayed a prolific astronomer till a very old age.

The Amazing Scientist Hertzsprung

A year after his appointment as deputy director in 1919 Hertzsprung became an extraordinary professor and received an honorary doctorate from Utrecht University in 1923. Actually he only lived for the experimental astronomy. Adriaan Blaauw, until his death in 2010, called him the most *absolute* scientist. In his own eyes he had devoted much too little time to real astronomy during De Sitter's stay in Arosa. Within the framework of the cooperation with the Union Observatory South Africa he stayed in Johannesburg for eighteen months (see the next chapter).

In the years 1926–1927 he stayed at Harvard College Observatory in Cambridge (Massachusetts, USA) and visited a dozen of other American observatories [ASL, 20-1-1927 from Hertzsprung]. Hertzsprung felt rather uneasy about the Americans thinking to be able *to buy* every good astronomer. The wish of the American observatories was understandable, because there was a lot of talent abroad which they needed. Shortly after the reorganization Leiden belonged, at least according to Schlesinger, to the six best observatories in the world [ASL, 20-1-1927 from Hertzsprung]. Hertzsprung discovered a peculiar photograph in Cambridge, with two takes, from the year 1900, on which the second one showed a much bigger imprint than the first one. He thought that it had to have been an event in our own solar system. He suggested to De Sitter a collision between two meteors, however improbable. He asked if De Sitter had a better idea. An answer has not been found yet. On plates, a day before and a day after, the object could not be seen. It is a serious riddle, Hertzsprung wrote [ASL, 1-3-1927]. In the years 1930–1931 he was again in Johannesburg for some time and in 1932 again in Cambridge. In those years Hertzsprung realized the ideal position he was in due to De Sitter: complete freedom to spend all his time to observations, and little organizational obligations and lectures [ASL, 25-4-1927 from Schlesinger]. Every summer holiday Hertzsprung spent a month in Denmark, but visited always a few observatories there, and also in Germany. The more theoretically minded students became students of De Sitter and Woltjer. But the experimentally minded students found in Hertzsprung the best teacher they could have wished for. In his period as deputy director Hertzsprung was advisor of Luyten (in 1921), Van Gent[12] (in 1932), Oosterhoff (in 1933) and Kuiper (in 1933). Actually Van den Bos (in 1925) was also a student of Hertzsprung, but De Sitter acted as supervisor, because Hertzsprung was in South Africa. In Van den Bos' preface he wrote as Hertzsprungs successor as Leiden astronomer in Johannesburg (Van den Bos 1925):

I will try not to lag too much behind the example set by you of tireless observing.

Luyten described Hertzsprung as one of the most brilliant, original and creative astronomers of his generation (Luyten 1987). Later De Sitter's son Aernout (1905–1944), Willem C. Martin (d. 1945), Wesselink, Lukas Plaut (1910–1984) and Jacobus Ferwerda would be doctor students of Hertzsprung. Aernout, guided by Hertzsprung, became an excellent photometrist. He was an assistant from 1928 until 1937 and received his doctor's degree in 1936 (De Sitter 1936). It is significant that Hertzspung realized only in 1930 that *Aernout* was his first name and not *Arnold*. [ASL, 28-6-1929, 7-1-1931]

---

[12]Hendrik van Gent (1900–1947) studied in Leiden and went in 1928 to the Union Observatory in Johannesburg.

Between Hertzsprung and De Sitter there was a discussion in relation to the conditions on the quality of a dissertation. Luyten had become a doctor too early, but the quality of the dissertations after him had to be high. The quantity of research had not or hardly to be less than the amount Van den Bos and Hins had done. That made it not easy for the future doctors, because the quality of Hins' and Van den Bos' work could hardly be equalled. Actually those two could easily have been awarded their degree earlier. Van Gent and Oosterhoff had enough material, but the young Aernout de Sitter had to show some more activity first, according to Hertzsprung [ASL, 14-1-1931]. Aernout's father agreed that the demands had to be high, but not too high [ASL, 4-2-1931]. Aernout seemed to become the victim of being his father's son. Perhaps Hertzsprung was a little afraid of giving him a too easy treatment, just because two very talented men had become a doctor. And probably just because of that he demanded even more [ASL, 25-2-1931]. In the end the two gentlemen agreed. It had to become simply a good dissertation [ASL, 15-4-1931 from Hertzsprung]. In this comparative discussion the students of De Sitter played no role, because they had written a theoretical dissertation. Hertzsprung stayed out of that.

We can deduce a lot about the discussions between De Sitter and Hertzsprung because there were many periods that they were not together in Leiden and had to communicate by letter. In the first period Hertzsprung was in Leiden and De Sitter stayed in Arosa, and when De Sitter was in Leiden, Hertzsprung was in South Afica and the United States regularly for long periods. In particular Hertzsprung was a prolific writer.

International Fame of the Dutch Astronomy

The international fame of Dutch astronomy in general and of the new Leiden Observatory in particular in the first half of the twentieth century was based on a number of elements: persons, publications, education, exchanges and organization. Part of these causes have been discussed already.

On a few of the leading actors the following remarks may be made. First there was the international fame of Kapteyn himself, in particular in the English-speaking world. By many he was considered the greatest astronomer of the late nineteenth and the early twentieth century. From 1885 onwards there had been his connection with Gill and through him with the English astronomy. And after Kapteyn had met Hale in St. Louis in 1904, there was also a tight friendly and scientific connection with America. There were the world's largest reflector on Mount Wilson and the largest refractor of the Yerkes Observatory. For Kapteyn it had been clear that cooperation with these observatories was necessary for his research into the structure of the Milky Way. Moreover, Kapteyn had played an important part in the planning of the new organization of the Leiden Observatory. De Sitter followed in Kapteyn's footsteps in making his name in England with his relativity articles. And Leiden had Hertzsprung, already at the start of his appointment in Leiden a great name as an astrophysicist. Later Oort would gain world-wide fame with his theory on the rotating Milky Way. Adding to that De Sitter had at his disposal a vast network of international contacts by his appointment as vice-president in 1922 and as president in 1925 of the International Astronomical Union.

Already a few years after De Sitter took over in 1919 Leiden Observatory was bubbling with activity. It became a beehive of astronomers. De Sitter and Hertzsprung succeeded in attracting and educating young and talented scientists, some of who went to work temporarily or permanently in observatories in America or South Africa. Besides foreign astronomers visited Leiden for a shorter or longer visit. Here follows a survey of the visits to observatories and meetings abroad and to Leiden from abroad. The data come from De Sitter's annual reports.

Guests from abroad			Leiden astronomers abroad		
1921	J. Jeans	Cambridge (GB)	20–21	W. Luyten	Greenwich
1921	A. Einstein	Berlin	1921	W. Luyten	Lick
1922	H. Shapley	Harvard	1920	C. Hins	Berlin
1923	R. Innes	Johannesburg	1921	J. Kapteyn	BA, Edinburgh
1923	H. Vanderlinden	Ukkel (Belgium)	1922	J. Oort	Yale
1923	A. Eddington	Cambridge (GB)	1922	W. de Sitter, E. Hertzsprung	IAU, Rome
1923	A. Einstein	Berlin			
1923	S. Rosseland	Copenhagen	1922	E. Hertzsprung	RAS, London
1924	R. Innes	Johannesburg	23–24	E. Hertzsprung	Johannesburg
1925	F. Schlesinger, B. Boss, A. Leuschner	USA	1923	W. de Sitter	Manchester, Edinburgh, London
25–26	S. Szeligowski	Poland	1923	W. van den Bos	Potsdam
1926	Sohon	USA	24–25	E. Hertzsprung	Johannesburg
1926	H. Shapley	Harvard	1925	W. de Sitter, J. Oort, J. Schilt, E. Hertzsprung, W. van den Bos	IAU, Cambridge (GB), Greenwich, RAS, London
1926	Ö. Bergstrand	Sweden			
1927	F. Stratton	Cambridge (GB)		W. van den Bos	Johannesburg
1928	IAU participants		1925	J. Schilt	Mount Wilson
1928	A. Nielsen	Aarhus	26–27	W. van den Bos	Johannesburg
1929	F. Schlesinger	Yale	1926	W/de Sitter, Hertzsprung	AG, Copenhagen
1929	R. Schorr	Hamburg			
1929	J. Hopmann	Bonn	26–27	Hertzsprung	Harvard e.a.
1929	O. Struve	Yerkes	1927	Doorn	Eclipse, Gällivare (Sweden)
1930	E. Rybka	Warsaw	1927	W. de Sitter	IUGG, Prague
1930	J. Mohr	Bratislava	1928	H. van Gent	Johannesburg
1931	F. Schlesinger	Yale	1928	G. Kuiper	Potsdam
1932	R. Aitken	Lick	1928	C. Hins	Cooke e.a., York
1932	H. Wood	Johannesburg	1929	H. van Gent	Johannesburg
1933	L. Gratton	Milaan	1929	W. de Sitter	BA, South Africa
1933	A. Van Hoof	Leuven	1929	E. Hertzsprung	RAS, Darwin Lecture (GB)
33–34	L. Plaut	Berlin	30–31	E. Hertzsprung	Johannesburg
33–34	K. Strand	Denmark	1930	H. an Gent	Johannesburg
1933	F. Schlesinger	Yale	30–31	A. de Sitter	Greenwich
			1930	W. de Sitter	IUGG, Stockholm
			1931	W. de Sitter	BA (GB)
			1931	W. de Sitter	RAS, Darwin Lecture (GB)
			31–32	W. de Sitter	USA, Canada
			31–32	C. Hins, G. van Herk	Kenia
			1932	E. Hertzsprung, H. Kluyver	IAU, Cambridge (USA)
			1932	J. Oort	IAU, Harvard, Perkins, e.a
			1932	J. Oort	Eclipse Expedition (Canada)
			1933	W. de Sitter	BA, Leicester (GB)
			1933	G. Kuiper	Lick
			1933	P. Oosterhoff	Harvard, Mount Wilson

Moreover, Einstein came as an extraordinary professor to Leiden for a number of years, each year for a couple of weeks, when he and De Sitter had much contact. Besides the shorter and longer scientific travels abroad De Sitter attended often the regular meetings of the Royal Astronomical Society. Each year there was at least one guest who stayed in Leiden for several months, did research and published about it. For a lot of observatories abroad Leiden was a good choice for their young astronomers to receive part of their education.

Then there was the choice for the English language as the scientific *lingua franca* for writing the publications. German and French had been the principal languages for publication in the nineteenth century. Kapteyn had made the choice for the English language already in 1900 for his *Publications of the Astronomical Laboratory at Groningen*. The Observatory in Leiden had published its own work in German from 1868 until 1915 in the *Annalen der Sternwarte in Leiden*. Even the name was German. De Sitter had changed the name in 1919 into the Dutch *Annalen van de Sterrewacht in Leiden*. In these Leiden Annals most publications were in English. Besides the own annals of each observatory in The Netherlands, mostly used to establish extensive observational sessions and their reductions, amongst Dutch astronomers there was demand for a magazine for the publication of smaller articles. As a result of the First World War the German *Astronomische Nachrichten* had ceased to be the obvious magazine for publication and the English preferred to keep their magazines exclusive. Hertzsprung proposed to start a Dutch magazine in The Netherlands. He considered the alternative to publish in the *Reports* of the Academy, like Easton,[13] as *an elegant grave* [ASL, 15-2-1921]. De Sitter considered it an attractive idea. Perhaps it should be initiated by the Dutch Astronomers Society. But sufficient subsidy was required and there had to be enough (international) subscriptions. And beforehand it had to be discussed with the English and American astronomers [ASL, 18-2-1921]. Pannekoek was of the opinion that a joint bulletin of the Dutch observatories would be a better idea [ASL, 12-9-1921]. In a bulletin only contributions of Dutch observatories would be published. Hertzsprung wrote to ask Bosscha if he was prepared to join in [ASL, 17-3-1921]. The Bosscha Observatory was positive. And a bulletin would not need to appear regularly, only when there were enough manuscripts. In the end it was decided to found the

---

[13]Cornelis Easton (1864–1929) was journalist and an amateur astronomer, making famous drawings of the Milky Way. He received an honorary doctor's degree from the University of Groningen, at the proposal of Kapteyn.

Bulletin of the Astronomical Institutes of The Netherlands (BAN). The Holland Society of Sciences (HMW) awarded, at the request of Nijland and De Sitter, a subsidy of 200 Dutch guilders yearly for a period of five years, in particular for the high printing costs. De Sitter and Nijland corresponded with the HMW about the subsidy in October and November 1921. With this Bulletin the results of the Dutch observatories achieved an international circulation, certainly in the English-speaking world. Leiden supplied the main part of the publications. For instance Volume II contained in 40 issues 76 shorter or longer articles: 59 from Leiden and 17 from the other observatories. Hertzsprung wrote 28 of the Leiden entries.

De Sitter was of the opinion that international experience was indispensable in a good education of an astronomer. His own excellent experience in South Africa undoubtedly lay at the basis of this view. Due to his extensive international network he achieved to obtain positions for the assistants of the Observatory in the twenties and early thirties. Pannekoek, also a man with much international experience, although in a completely different field, shared the view that young Dutch astronomers must certainly not stay in one place, but had to visit other observatories to prevent specialization [ASL, 13-7-1923]. Besides that the education of astronomers in the United States did not keep pace with the demand. The big telescopes needed a substantial and qualified staff. Quantitatively as well as qualitatively there were shortages. There was a logical place to recruit astronomers: The Netherlands and in particular Leiden.

Shapley wrote to De Sitter in 1928 that he remembered De Sitter claiming as a principal task to educate and export highly qualified young astronomers [ASL, 22-12-1928]. Shapley once said at a conference to a neighbour, who introduced himself as a Dutchman, the legendary words:

Ah, that is the country where they grow tulips and astronomers for export.

At least part of it were De Sitter's own words. As a result of this Leiden policy and the American need for good astronomers for this fast growing branch of science, a flow of astronomers started, who temporarily or permanently went to work abroad, mostly in the United States. Many of them achieved a director's position. Not only Leiden but also Groningen and Utrecht were suppliers. Here follows a list with astronomers who settled permanently abroad. The list mentions name, advisor and place of doctoral degree, year of departure and foreign observatories and universities.

Export of Dutch astronomers (Baneke 2015)	To
Bart Bok, Van Rhijn, Groningen, 1929:	Harvard, Mount Stromlo (Aus), Steward (Tucson, USA)
Willem van den Bos, De Sitter, Leiden, 1925:	Union Observatory (Johannesburg)
Dirk Brouwer, De Sitter, Leiden, 1927:	Berkeley, Yale (USA)
Paul ten Bruggencate, Von Seeliger, München:	Bosscha, Mount Wilson, Potsdam, Göttingen
Peter van de Kamp, doctoral degree in Califormia and with Van Rhijn, 1923	McCormick, Lick, Sproul (Swarthmore, USA)
Gerard Kuiper, Hertzsprung, Leiden, 1933:	Lick, Yerkes, McDonald (Austin, USA)
Willem Luyten, Hertzsprung, Leiden, 1921:	Lick, Harvard, Minnesota (USA)
Adriaan van Maanen, Nijland, Utrecht, 1911:	Yerkes, Mount Wilson
Dirk Reuijl, Nijland, Utrecht, 1928:	McCormick (USA)
Jan Schilt, Van Rhijn, Groningen, 1926:	Yale, Columbia (USA)

Besides these astronomers, who went permanently abroad, there were many who went for shorter or longer periods to the USA for work or study, amongst whom Kapteyn, Julius, Hertzsprung, Van Rhijn, Van der Bilt, Zanstra, Oort, Pannekoek, Ten Bruggencate, De Sitter, Minnaert and Oosterhoff.

De Sitter, Schlesinger, Shapley and Hertzsprung did a lot of work during the twenties to find and organize places in the USA for Dutch astronomical talents. They formed a successful group of *brokers*. From the correspondence of De Sitter one can see how much effort he made to find jobs abroad for the young astronomers. The gratitude for the Leiden education, in which the trio De Sitter, Hertzsprung and Woltjer played the main role, appears from the letters to the home front. Not seldom also sincere gratitude may be read for De Sitter's interest and support in personal matters. Hertzprung gave an important contribution during his stays in the USA.

Oort was one of the first. De Sitter and Schlesinger arranged a job for Oort, just one year in possession of his master's degree, with Schlesinger at Yale. Oort, being afraid that he only would have to make observations, was reassured by De Sitter. Certainly there would be time left to do some work of his own [ASL, 22-5, 7-6, 22-6, 1922]. De Sitter already saw, as Kapteyn had done even earlier, Oort's vast talents. Schlesinger wished to give him a broad exercise [ASL, 9-6-1922]. He was glad with his young assistant and wrote to De Sitter completely agreeing with him about his great chances for the future. Offers were already made to Oort then, but De Sitter preferred to have him back in Leiden. He offered him a job as a curator at the Observatory, which Oort accepted. Schlesinger let him go reluctantly, but was also of the opinion that Oort must not only become an observer, but had to continue the theoretical statistical work of Kapteyn too. According to Schlesinger Oort had occupied an important place at Yale and would achieve great successes in *our*

Fig. 11.4 Frank Schlesinger

science. At the end of 1924 Schlesinger sailed to Europe and would also visit Leiden and the De Sitters. He wished to get an impression about Schilt [ASL, 14-2, 24-3-1923, 15-10-1924]. De Sitter and the faculty managed to create a lecturership for Oort in 1929. De Sitter would consider it of *national importance* to prevent the *insurmountable disaster* of a possible departure of Oort to a professorship at Harvard [ASL, 12-1-1929 to BG]. The faculty also wrote to the Board of Governors [ASL, 28-1-1929]. The Minister wrote to the Board of Governors and Oort could stay [ASL, 13-2-1929]. In 1930 Oort again faced a difficult choice. He had received an offer from Columbia University in New York. Schlesinger wrote to De Sitter that he thought Oort's decision would probably be to the advantage of Leiden again [ASL, 20-3-1930]. A year later Schilt went to Columbia University.

De Sitter, Hale [ASL, 28-1-1925 from Hale] and Schlesinger also did their best for Schilt, in Leiden since 1922. De Sitter tried to obtain a fellowship for Schilt from the International Education Board (part of the Rockefeller Foundation). That was successful with recommendations, amongst others from Van Rhijn. First Schilt visited the general assembly of the International Astronomical Union in Cambridge (UK) in 1925. Later that year, at the end of August, he left for the USA. From Mount Wilson, where he, at the request of De Sitter, had to learn about large telescopes with Hale's successor W.S. Adams (1876–1956), Schilt wrote to De Sitter. He found it a disgrace that his wife and he had been detained on Ellis Island (New York) in connection with their visa. Probably I am going to do something in the field of Hubble's work, he wrote. Adams was also interested in thermopile photometry and perhaps wished to order a Schilt photometer. A year later Schlesinger offered him a job, which Schilt accepted, but the Schilts could not stay for beaurocratic reasons. They had to go back first, to enter the USA again later. De Sitter wrote extensively to the envoy of The Netherlands in Washington to get his support. He wrote:

> Going back is a disaster. There are no young American astronomers. Thus it is necessary to import young scientists from Europe. It is a postwar phenomenon. Young Americans cannot be found anymore for the badly paid scientific jobs. It is a waste that he has to go to Europe and back again for beaurocratic reasons.

Schlesinger tried it too:

> Kapteyn, the greatest astronomer of the world in his day, called him one of his two most brilliant students. If we must make him Assistant Professor, we'll do that.

The other student Kapteyn referred to was no doubt Oort. At last a solution was found. The Schilts had to leave the USA and stay for a short time in Ottawa in Canada, after which they could enter the USA again [ASL, Correspondence of Schilt and Schlesinger, 1925 and 1926]. Schilt thanked De Sitter for the large contribution he had made to his career [ASL, 1-4-1927].

Jan Schilt (1894–1982)
Jan Schilt studied in Utrecht and Groningen (under Kapteyn) and came to Leiden in 1922. He received his doctor's degree with Van Rhijn in 1924. The title of his dissertation was *On a Thermo-Electric Method of Measuring Photographic Magnitudes*. Based on this method he designed and built the later called Schilt Photometer. With this apparatus the luminosity of stars on photographic plates could be measured. The intensity of a through-going light beam was a measure of the intensity on that spot of the plate. The intensity was measured with a thermocouple with a galvanometer, designed by the later Nobel prize winner F. Zernike (1888–1966).

Schilt went to Mount Wilson with a grant in 1925. The next year he went to Schlesinger at Yale Observatory, where he worked on photographic determination of parallaxes and proper motions. Later, in 1931, he was appointed professor at Columbia University in New York. He had an extensive knowledge of the flora (while a student he had taught even biology) and was a great lover of classical music. Schilt had an independent mind. After the launch of the first artificial satellite in 1957, the Russian Sputnik, he proposed to undertake a space travel to one of the two small moons of Mars. And already in 1945, shortly after the two atomic bombs on Japan, he wrote a letter to the editor of The New York Times, in which he wrote about the necessity of international control on nuclear arms. Privately, the Schilts and the Oorts were very good friends.

Another year later, in 1927, Brouwer received a fellowship at Yale. Already in 1928 Schlesinger wrote to De Sitter that Brouwer would stay permanently as an assistant of Brown. Brouwer would make a great career in celestial mechanics. Schlesinger informed De Sitter regularly about his *young boys* [ASL, 28-7, 10-10-1927, 29-3-1928].

Oosterhoff was another talented young astronomer. He felt attracted to the practical astronomy. Van den Bos and Oosterhoff became the two students of Hertzsprung he valued most. Oosterhoff became Hertzsprung's right-hand man. According to Hertzsprung the main aim was to leave as many observations as possible from the present epoch for the future astronomers. Oosterhoff did not speculate on possible theories, rarely he ventured an interpretation. His contribution to the data of variable stars was enormous (Oort 1978). Hertzsprung wrote a letter of recommendation to The Rockefeller Foundation and Oosterhoff received a fellowship. Both Adams of Mount Wilson and Shapley of Harvard were keen to give Oosterhoff a job. He worked at both Observatories. Shapley wrote to De Sitter at the end of 1933 that Oosterhoff would be wise not to work so hard, but to talk, look around and think more. But he added:

One cannot prevent a product of the Leiden Observatory to work.

[ASL, 17-11-1933]

The next of the export astronomers was Bartholomeus (Bart) Bok (1906–1983). He studied in Leiden and later in Groningen. Bok was a student assistant in the reception committee at the general assembly of the International Astronomical Union in Leiden in 1928. There he met Shapley and his future wife Priscilla Fairfield (1896–1975), an American professor of astronomy. The next year Shapley invited him to come to Harvard (Millman 1984). Before he left for America he wrote to De Sitter that, in particular by his support, he had had a wonderful student period. He thanked him for his help with personal ups and downs and his encouragement and help with his appointment in Groningen. He would uphold the great reputation of Leiden Observatory in America [ASL, 1929]. A few years later, in 1932, he would receive his doctor's degree in Groningen with Van Rhijn as his advisor on *A study of the Eta Carinae Region*. He worked closely together with his wife, first at Harvard from 1929 until 1957, and then at Mount Stromlo Observatory in Canberra (Australia) from 1957 until 1966.

The last in the series of export astronomers was Kuiper, who received a job at the Lick Observatory on Mount Hamilton in California (USA). From 1933 he wrote a number of times to De Sitter about his experiences at Lick. He wrote until September 1934, two months before De Sitter's death. He confirmed De Sitter's prediction that he would find Lick an agreeable place. You know the place, he wrote, because De Sitter had visited Lick on his America trip at the end of 1931. Later he was a lot less positive. From the Dutch East Indies Kuiper received an invitation from Voûte, director of the Bosscha Observatory in Lembang, to come and work there. He thought about it seriously, because there was a fair amount of protection at Lick. Foreigners were not particularly liked. When I go to Lembang, he wrote, it will be good for many people. At that Kuiper was not impressed by the level of the American astronomers. Some of them were very mediocre. In his letters he showed a lot of interest in the fortunes of Leiden Observatory. Ehrenfest's death was a great blow to him. And he would like to see the film about the Observatory made at the 25-years' jubilee of De Sitter's professorship.[14] He thanked De Sitter for the photograph of the portrait of De Sitter painted by the Dutch painter Jan Sluijters, on which he found the spiral nebula in the star constellation The Hunting Dogs (*Canes Venatici*) very suggestive [ASL, Kuiper correspondence]. But Kuiper stayed in the USA and continued his career at the observatories of Yerkes and McDonald.

From all descriptions above we can see the enormous amount of energy De Sitter and Hertzsprung did put in the internationally oriented education of their students. With the help of Schlesinger, Shapley and others they managed to provide their students with jobs at excellent observatories. From the number of astronomers that stayed in the USA in the end we can deduce the profound influence the Dutch school of astronomy had on the advancement of astronomy in the USA.

References

Baneke, D.M. (2015). *De Ontdekkers van de Hemel, De Nederlandse sterrenkunde in de twintigste eeuw* (*The discoverers of the heaven, Dutch astronomy in the 20th century*), Prometheus - Bert Bakker, Amsterdam, 2015.

De Sitter, A. (1936). *Fotovisuele fotometrie van sterren tot 8.^{m}0 benoorden +80° declinatie* (*Photovisual photometry of stars up to 8 ^{m}.0 north of +80° declination*), Luctor et Emergo, Leiden, 1936.

De Sitter, W. (1926). 'Oration of the Rector Dr. W. de Sitter on the 351th *dies natalis* of Leiden University on 8 February 1926: The unity of science'; *Annual of Leiden University 1926*, S.C. van Doesburgh, Leiden, 1926. (Dutch: 'Oratie van den Rector Magnificus Dr. W. de Sitter op den 351sten dies natalis der Leidsche Hoogeschool 8 februari 1926: De eenheid der wetenschap'; Jaarboek der Rijksuniversiteit te Leiden 1926'.)

De Visser, J.Th. (1924). *Speech at the Inauguration of the Reorganized Observatory in Leiden by His Excellency the Minister of Education, Arts and Sciences* (Dutch: *Toespraak bij de Inwijding van de Gereorganiseerde Sterrewacht te Leiden door Zijne Excellentie den Minister van Onderwijs, Kunsten en Wetenschappen op 18 September 1924*), Joh. Enschedeé en Zonen, Haarlem, 1924. Present in ASL.

Eddington, A.S. (1934). 'Obituary'; *Nature*, CXXXIV, pp. 924–925, December 15, 1934.

---

[14]The film of about ten minutes about the Observatory is an amateur film from 1933. It seemed to be lost, but the author was able to find it in the archives of the Netherlands Institute for Sound and Vision in Hilversum (province of Noord-Holland). The film has to be restored, but a single unrestored copy could be made and was shown at the book presentation of the smaller version of this biography in Dutch in Leiden in 2009. It contains the only moving images of De Sitter (at his desk, while opening his letters).

Hins, C.H. (1925). *Inleiding tot een Catalogus van Plaatsen en Eigenbewegingen van 1533 Roode Sterren*; N.V. Electrische Drukkerij v.h. Taconis, Leiden, 1925. (*Introduction to a Catalogue of Positions and Proper Motions of 1533 Red Stars.*)

Luyten, W.J. (1987). *My first 72 years of astronomical research*, Privately published, 1987.

Millman, P.M. (1984). 'In Memoriam Bart Bok'; *Journal of the Royal Astronomical Society of Canada*, 78, 1, pp. 3–7, 1984.

Oort, J.H. (1951). 'Obituary Dr. C.H. Hins'; *The Observatory*, pp. 243–244, December 1951.

Oort, J.H. (1978). 'In Memoriam P.Th. Oosterhoff'; *Zenit*, 5, pp. 358–359, 1978.

Van den Bos, W.H. (1925). *Micrometermetingen van Dubbelsterren* (*Micrometer Measurements of Double Stars*), Eduard IJdo, Leiden, 1925.

Chapter 12
International Business

Johannesburg has the climate and the instruments, Leiden has the people and the brains.

Innes to De Sitter, 1923

The International Research Council

After preliminary conferences in 1918 in London and Paris the founding meetings of the International Research Council (IRC) were held. After the end of the First World War the enmity of the allied countries towards the central powers, in particular Germany, was so intense that the old international scientific structure could no longer be maintained. A new structure without Germany, Austria-Hungary, Bulgaria and the Ottoman Empire was formed. The first board consisted of the French mathematician C. E. Picard (1856–1941), the American Hale, the Belgian astronomer G. Lecointe (1869–1929) and the Italian mathematician V. Volterra (1860–1940). The English mathematician and physicist A. Schuster (1851–1934; German born) became its secretary.

For several branches of science Unions were founded, the first of which was the International Astronomical Union (IAU) in 1919.[1] See further in this chapter (Blaauw 1994).

The Netherlands had been neutral during the First World War and were as a consequence not involved in the founding meeting of the IRC. But having been neutral they immediately received an invitation to enter. In The Netherlands the debate on joining the new international organization IRC was pursued immediately after the end of the war in the meetings of the Academy of Sciences. Kapteyn opposed fiercely joining the IRC and with a number of like-minded colleagues he tried to map out a different route. With a League of Nations, in the process of being founded, there could not stay two camps, in the opinion of Kapteyn.[2] The Netherlands were invited to join the IRC in September and after long discussions on

---

[1]AIP, Minutes of the founding meeting in Brussels, 1919.
[2]NAW2, Report of the meeting of 25 January, 1919.

© Springer Nature Switzerland AG 2018
J. Guichelaar, *Willem de Sitter*, Springer Biographies,
https://doi.org/10.1007/978-3-319-98337-0_12

7 October in a combined meeting of both departments of the Academy the decision to join was made. Secretary P. Zeeman (1865–1943) informed the IRC by letter of 10 November 1919. But normalization of the relation with the central powers was necessary as soon as possible. Moreover, The Netherlands reserved their right for individual scientists to keep in contact with all other scientists in all countries [NAW1, 10-11-1919, NAW to IRC]. For Kapteyn this was a bitter experience, who was hardly present at the meetings anymore after this decision. Two letters from the correspondence in those days between Hale and Kapteyn give a good impression of the emotional discussions that were pursued in many countries. After the decision Kapteyn wrote an emotional letter to his old friend, the instigator and one of the most prominent advocates of the new structure. He wrote that he had fought, but had lost, and that he could imagine that Hale perhaps wished to end their close cooperation. Kapteyn would understand that, although he would remain grateful for all that Hale had done for him. He also informed Hale that he would accept no function whatsoever in the new astronomical union. It must have been a difficult letter for him. Hale wrote to Kapteyn and said that not until after a visit in 1916 to France the barbaric atrocities of the Germans had struck him, and that it would take years before a real cooperation would be possible again. He was very sorry that Kapteyn and he had such different opinions.[3]

The Dutch Academy Committee for International Scientific Cooperation[4]

The department of mathematics and physics of the Academy decided not to join each scientific union of the IRC, but to leave it to the representatives of the separate sciences. In order to still achieve some unity in the Dutch viewpoints on the different unions the Committee for International Scientific Cooperation was founded (WIS). The first meeting was held on 24 February 1923 with a representative from each branch of science.[5] The board of the department of mathematics and physics of the Academy were ex officio members of the Committee. De Sitter was the member representing astronomy and he could inform the Committee that the Dutch Astronomers Society (NAC), founded in 1918, had sent a delegation to the first general assembly of the International Astronomical Union in Rome in 1922 and that a lot of important scientific work had been done there. In the WIS-Committee the international cooperation was analysed in the ensuing years and the course was set. The entry of the Germans was not yet settled. One of the aims of the Committee was

---

[3]Archives of the California Institute of Technology, the G. E. Hale papers, Role 41, 12 October, from Kapteyn to Hale, 1919, 15 October, from Hale to Kapteyn, 1919. Copies are present in the IAU archive in Paris.

[4]Dutch: Wetenschappelijke Internationale Samenwerkingscommissie (WIS).

[5]NAW1, Reports of the WIS-Committee.

that the affiliation to the IRC of the central powers would be made possible at the next meeting of the IRC in 1925. Affiliation to the IRC was a *conditio sine qua non* for joining the International Astronomical Union. Lorentz, member of the Committee representing the physicists, remarked that not being a member of the League of Nations was no obstacle to joining the IRC and the scientific unions. A second aim in the plan of activities was the formation of national councils for each science. De Sitter tried to get the NAC appointed by the Academy as the council for astronomy, but that was a business for the astronomers themselves, according to the opinion of the chairman, the botanist F. Went (1863–1935). In 1924 a growing number of pleas could already be heard for the acceptance of the central powers. Schuster and Hale understood the pressure, but they considered the time not yet ripe for it. De Sitter and others saw future difficulties by the opinions of the IRC. They argued in favour of a more independent position for the unions. In his opinion the IRC could then become extinct. At the end of 1924 De Sitter remarked that he planned, after an idea of the NAC, to take the initiative to elicit from the general assembly of the International Astronomical Union in Cambridge (GB) in 1925 the statement to make changes in the statutes of the IRC. It would even be better if the initiative would come from the board of the IAU itself. In the Committee meeting of 28 February 1925 De Sitter reported about his activities. He had written letters to all board members of the IAU and had received mixed answers. England was positive, De Sitter had heard during a visit, but wished no conflict with France. The Americans also wished to restore the complete international cooperation. Through his extensive network De Sitter performed a lot of lobby work in those years in order to restore the unity in the scientific world. In April the WIS-Committee discussed the agenda of the coming meeting of the IRC, of which a proposal of The Netherlands and Denmark to adapt the statutes was one of the subjects. De Sitter announced that he had received positive communications from the American and Japanese deputy chairmen of the IAU. He pointed again to the desirability to make the unions independent from the IRC. After the July meeting of the IRC in Brussels the Committee convened again. A proposal to change the statutes of the IRC had been put to the vote in Brussels. The proposal was adopted by a simple majority, but Picard said that a two-thirds majority was necessary just because it was a change in the statutes. So in Picard's view the proposal was rejected (Blaauw 1994). It created a lot of consternation. The rejection in Brussels had put De Sitter in an awkward position in the general assembly of the IAU in Cambridge a week later. On behalf of the Dutch government he would offer to organize the next general assembly in Leiden in 1928. Point of departure for the government was that all countries, including the central powers, could then be members. The offer was still made and accepted. For a short while the IAU had entertained the thought to step out of the IRC completely. The Italians and the Americans had declared not to attend meetings, if not every country was allowed to participate. In the WIS-Committee proposals were also heard to step out of the IRC altogether. But that would mean stepping out of the IAU too, and for that reason De Sitter opposed these ideas. De Sitter wished an independent IAU, but not a structure The Netherlands had to quit. The astronomical interests had to come first. In the June meeting in 1926 of the IRC the definite decision to change the statutes had to be taken. The Royal Society would present the proposals. But part of the proposal was explicitly that the membership of the IRC was a necessary condition for membership of a union. At last, in the June meeting of the IRC, where De Sitter was present as substitute of the English astronomer Dyson, the decision was made to implement the statute changes. The strongest opponent, chairman Picard, had put the proposal on the agenda. A next problem would be, the door being ajar then, to induce Germany to join the IRC, a condition for membership of the IAU. After many years of exclusion a lot of Germans were not particularly motivated. In the end Germany would join the IAU officially only after the Second World War in 1951.

International Astronomical Union

The International Astromical Union and the International Research Council were founded at the same time in Brussels in 1919. The first discussions were about the forming of sections, each of them managing a group of subjects. But in the end it was decided to found a separate committee for each subject: 32 in total. Eddington became chairman of Committee 1 about Relativity, then a hot topic, and Sampson of Committee 31 about Time. The neutral Netherlands were not yet member of the IAU, so no Committee was assigned to a Dutch astronomer. The French astronomer E. Baillaud (1848–1934) was the first president of the IAU, and the English astronomer A. Fowler (1868–1940) became general secretary. Further there were four vice-presidents: the American astronomer W. Campbell (1862–1938), Dyson, Lecointe and the Italian Ricco (1844–1919).The budget for the first complete year 1920 amounted to £981, furnished by the first seven members of the IAU: Belgium, Canada, England, France, Greece, Japan and the United States. The main part of the money went to the Bureau de l'Heure in Paris, which had been added to the Committee 31 (Time) at the start of the IAU in 1919. In the year 1921 the budget was £ 1350, due to two new members: Italy and Mexico. In 1919 De Sitter's time had completely been taken up by the start of the reorganization of the Observatory and at the end of that year he went to Arosa for 18 months. Only when he had returned, the IAU came into view again. After De Sitter had reported himself back to Dyson, then vice-president of the IAU, the last wrote to hope that Arosa had made him completely healthy again. He was also glad that The Netherlands would probably join the IAU and that De Sitter had to use all Schuster's knowledge to wind up all formalities. They would probably meet the following year in Rome, at the first general assembly of the IAU [RGO, 8/88, 2-6-1921].

Italy had already made the offer in 1919 to organize the first general assembly of the IAU.

De Sitter and Hertzsprung proposed to join the IAU in the Academy meeting on 24 September 1921 [NAW1, 24-9-1921]. De Sitter wrote that The Netherlands could already be considered as a great astronomical power [NAW1, 22-8-1921]. The decision was taken and The Netherlands could send a delegation to Rome.

First presents of the IAU

É. B. Baillaud	1919–1922
W. W. Campbell	1922–1925
W. de Sitter	1925–1928
F. W. Dyson	1928–1932
F. Schlesinger	1932–1935

The First General Assembly of the IAU in Rome in 1922

The IAU and the International Union of Geodesy and Geophysics met at the same time in Rome. As representatives of The Netherlands the Dutch astronomer J. de Vos van Steenwijk (1889–1978), De Sitter, Hertzsprung, Nijland and father Stein S. J. travelled to Rome. Oort was also in Rome and De Sitter provided him entrance to the meetings. After the assembly Schlesinger immediately wrote to De Sitter that he was impressed by Oort. De Sitter had recommended him and Schlesinger looked forward to Oort's coming to America [ASL, 9-6-1922]. The opening ceremony was on 2 May in the presence of King Victor Emanuel III and cardinal Pietro Maffi (1858–1931), who had taken part in the conclave that chose the new pope Pius XI a few months earlier. His presence was not surprising, because he was an active scientist. In a plenary meeting on 4 May, in a discussion about the expenses connected with the number of languages in which the decisions had to be printed, De Sitter proposed the compromise to print only the decisions of the general assembly in two languages: English and French. This proposal was carried. There followed a number of days with committee meetings. Committee 1 about Relativity was of the opinion that progress could only be made by individual activities and that international cooperation in this field was not useful. At the last plenary meeting on 10 May Cambridge (GB) was accepted as location for the next general assembly. As the final item of the agenda the new board was chosen, in which De Sitter became one of the vice-presidents and Campbell the second president.[6] Fowler stayed general secretary. It was the first appointment from a neutral country. De Sitter's theoretical achievements and his reorganization of the Leiden Observatory had not escaped his colleagues' attention. There was an audience with the pope after the closing of the general assembly on that day.[7] The De Sitters preferred to have lunch with the Schlesingers (De Sitter-Suermondt 1922). In the last meeting the new members of the committees had been appointed. De Sitter was

---

[6]The other members of the general board were: V. Cerulli (1859–1927) from Italy, H. Deslandres (1853–1848) from France, S. Hirayama (1868–1945) from Japan and S. S. Hough from South Africa.

[7]'General Notes'; *Popular Astronomy*, 30, pp. 449–450, 1922.

appointed in as many as five committees: Relativity (1), Dynamical astronomy and astronomical tables (7), Meridian astronomy and study of refraction (8), Stellar parallaxes (24) and Time (31). There were few colleagues taking so much on their plates. On this trip De Sitter was accompanied by his wife and both used the opportunity to visit the touristic attractions. In particular the gardens of the Villa Borghese, which they visited together with Oort, made a big impression. All days were fruitful with many astronomical contacts (De Sitter-Suermondt 1922).

On their way home De Sitter and his wife stayed in Florence for some time, to recover from the fatigues. From Florence De Sitter wrote to the president of the Board of Governors of Leiden University De Gijselaar, that he had been impressed by the recognition The Netherlands had received, and by the fact that a Dutch astronomer had been elected one of the vice-presidents: De Sitter. Of course it was now relevant to comply with all international research arrangements. Thus all agreed amounts for the Observatory had to be part of the 1923 budget. Being too tired himself to come immediately to Leiden de Sitter asked De Gijselaar to exert his influence with the Minister to get all the amounts in the budget. Further he asked another two weeks of leave [AUL, BG, 14-5-1922]. After his appointment as director of the Observatory, the one as vice-president of the IAU was the next step in De Sitter's career.

Due to the fact that De Sitter did not return to Leiden immediately, he missed a meeting of the NAC. There, Shapley, having gone straight from Rome to Leiden, would lecture on his ideas about the size of the Milky Way or universe (in those years for many astronomers there was no difference between the two). It certainly was a subject that would have interested De Sitter. Some colleagues, like Kapteyn and Easton, were sceptical about the size calculated by Shapley. Pannekoek wrote to De Sitter after the meeting that there had been transparencies, but unfortunately little discussion [ASL, 28-1, 21-2, 14-5, 22-5, 23-5, 7-6-1922]. The Dutch delegations to the general assemblies would officially be sent by the NAC. In later years the conviction grew that actually a private society was not the appropriate body. After consultation of the NAC Oort, then general secretary of the IAU, directed a request to the Academy of Sciences in 1939 to take over this task [NAW1, 19-12-1939].

The Second General Assembly of the IAU in Cambridge (GB) in 1925

The general assembly of 1925 in Cambridge was visited by a Dutch delegation of eleven astronomers, four of them from Leiden: De Sitter, Hertzsprung, Oort and Schilt. England had a delegation of 85, the USA of 23 and France of 14. With eleven members The Netherlands were the fourth in magnitude of the nineteen countries taking part, a *great power* in De Sitter's words.

During the first plenary meeting on 15 July there was a long discussion about a new definition of the Greenwich Time, which had to change date at midnight and not at noon. The astronomers were against the plan, because they had to change the date during their nightly observations. In the end it annoyed De Sitter and he proposed that the general public and the astronomers would leave each other alone. Both parties had to use their own time. The Committee 1 about Relativity did not convene and the general board proposed to abolish the Committee, as could have been expected already in 1922. For the purpose of the research into the structure of the Milky Way research programmes were agreed on, concerning variable Cepheid stars, star clusters and large nebulae. Hertzsprung presented his results on a possible connection between the periods and the forms of the light curves of Cepheid variables, the result of his work in Johannesburg. Eddington spoke about stars with densities of 60,000 times that of water (Henroteau 1925). De Sitter managed to get a proposal accepted in the Committee 8 for Meridian astronomy to grant three times £250 for the coming years for his plan to remove systematic errors from the declination measurements with the help of azimuth measurements. The proposal was carried by the plenary meeting. In the last meeting of the general assembly on 22 July the invitation of the Dutch government to organize the next general assembly in Leiden in 1928 was unanimously accepted. De Sitter was chosen as the next president and the English astronomer Colonel F. Stratton (1881–1960) as general secretary.[8] The choice for a president from a formal neutral country was again a step forward to a complete international cooperation. During the preparatory stages to the general assembly in Leiden de Sitter would be very active to restore a complete international cooperation.

The general assembly in Cambridge had a second recognition in store for him. He received an honorary doctorate from the University of Cambridge on 21 July. Four others received an honorary doctorate too: Baillaud and Campbell, first and second president of the IAU, the Japanese physicist H. Nagaoka (1865–1950) and Schlesinger. At lunch the *young* doctors, in a red gown, were addressed in a humorous manner. In the afternoon there was the official part with a speech for each doctor in Latin, and in the evening, during dinner in the Hall of Trinity College, the *loving cup* went round for everybody to take a sip. In his speech of gratitude De Sitter praised the radiating scientific light from Cambridge.

It would be difficult to equal the magnificent receptions of the English in The Netherlands in 1928, the Dutch daily paper *NRC*[9] wrote on 27 July.

---

[8]The other members of the general board were: Eddington, Schlesinger, Cerulli, Deslandres, Hirayama, all vice-presidents.

[9]Dutch: *Nieuwe Rotterdamse Courant.*

The Third General Assembly of the IAU in Leiden in 1928

The German Astronomische Gesellschaft (AG), with many international members, did not abandon its international position just like that. In the period leading up to the assembly in Leiden de Sitter had a lot of problems with the German organization. An aggressive article of G. Struve (1886–1933) from Berlin, a member of the extensive Russian-German-American Struve family of astronomers, had appeared in the German press in 1927, in which the in his eyes disgraceful behaviour of the French in particular was denounced. For the Germans the AG was sufficient, they did not need the IAU. De Sitter wondered if he as president had to write to the Germans and asked advice from Schlesinger and Lorentz [AL, 23-8-1927]. For a good functioning of the IAU the Germans were not necessary in the eyes of De Sitter. Schlesinger was not surprised about the attitude of the Germans. He advised De Sitter against writing directly. They could always join later [ASL, 6-9-1927]. In the 1922 assembly an advice to the IRC was adopted that astronomers from countries that were allowed to become members could be invited by the president. The IRC extended this presidential competence also to the central powers in 1926 (Blaauw 1994). But this had not yet been included in the IAU statutes. According to general secretary Stratton it was formally not yet possible for De Sitter to invite Germans, but Schlesinger said that this could be done immediately in 1928. So De Sitter could go ahead with inviting Germans to the Leiden assembly. De Sitter planned to invite German and Russian astronomers. There were already contacts with the Russians to join the IAU. De Sitter corresponded with Schlesinger on the subject [ASL, 6-10, 25-10, 7-11-1927]. De Sitter also had corresponded with Shapley about his view on the invitations. The last had heard from Luyten the rumour that H. Ludendorff (1873–1941) and P. Guthnik (1879–1947) would not come, but he hoped that they would not be so stupid to turn down an invitation. He was glad that De Sitter would invite a great number of German astronomers. In his letters Shapley proposed already to plan the fourth general assembly in Cambridge (USA), not in 1931 but in 1932, because in that year there would be a solar eclipse in the region of Québec [ASL, 23-7, 6-9, 10-10, 17-10-1927, 23-1-1928].

   In the end delegations from twenty-nine countries came to Leiden with a total of 250 delegates. Germany appeared with a sizeable delegation of fourteen members. All De Sitter's diplomatic activities had been successful. He still had in his mind the beautiful ceremony with five honorary doctorates in Cambridge in 1925 and decided to steer the same course. He managed to obtain the approval of the faculty, which sent a letter to the Senate with the proposal to honour Deslandres, Dyson, Eddington, F. Küstner (1856-1936) and Shapley with an honorary doctorate, on the occasion of the general assembly of the IAU. On the original proposal was written that the letter had been withdrawn from discussion in the Senate at the request of the faculty. This was done after the proposal had been discussed in the smaller group of Rector and Assessors (for the day-to-day running of the Senate), who probably had expected problems with such a large number of doctorates and had contacted the faculty [AUL, Faculty, 28-2-1928 to the Senate]. The letter was exchanged for another one with the proposal to bestow an honorary doctorate upon the German Küstner and the Frenchman Deslandres, the greatest adversaries in the war. These awards would confirm the restoration of the international cooperation after the First World War. In this exceptional situation it would certainly not create a precedent. The Senate acted upon the request and decided that the honorary doctorates would be presented on 10 July, in a ceremonious Senate meeting. This would be followed by a tea in the Botanical Garden, offered by the Senate to the general assembly [AUL, Faculty, 7-6-1928 from the Senate]. The organization was mainly in the hands of general secretary Stratton, supported by the local secretary Hins. The official opening of the general assembly was performed by the Minister of Education Mr. M. Waszink, in the Knights' Hall[10] on 5 July. A number of extra trains were available. In the evening there was a reception by the city council. The first plenary meeting was held on the next day in the City Auditorium[11] of Leiden. De Sitter addressed the German delegation separately in German. On 10 July the honorary doctorates were conferred to Deslandres, for his spectroscopic work, and to Küstner, for his work on fundamental astronomy. Guthnik accepted the certificate in the name of Küstner, who could not be present owing to ill health. With this doctor's certificate an old wish of Küstner's friend Kapteyn was fulfilled.[12] The work of the Committees also took a number of sessions. In Committee 8, for Meridian astronomy, Hins gave a survey of the current state of affairs concerning the fundamental declinations. The azimuth instrument would be delivered by the Cooke factory within a few months.[13] Van Rhijn announced on 12 July that the biography of Kapteyn, written by his daughter Henriëtte Hertzsprung-Kapteyn, had been published very recently. During the closing meeting Cambridge (USA) was assigned as location for the next general assembly. Deslandres hoped that thereafter Paris would be chosen. Good care had been taken for a programme with excursions, amongst others to the Southern Sea Works.[14] [15] The new general board was chosen and Dyson became the next president.[16] Stratton stayed general secretary. At the end of the meeting many words of gratitude were spoken, amongst others to the indefatigable Hins and the chairman of the Leiden Students' Union. Mrs De Sitter was thanked too, as chairwoman of the Ladies' Committee. Schlesinger emphasized how De Sitter had steered the IAU through a difficult period of three years. De Sitter was moved by so much gratitude. He received a minutes' long standing ovation. It

---

[10]Dutch: Ridderzaal. In the Knights' Hall in The Hague the combined meeting of the two houses of parliament is held, during which the King/Queen delivers the King's/Queen's speech.

[11]Dutch: Stadsgehoorzaal.

[12]Daily paper NRC, 10 July 1928.

[13]With this instrument Hins and Van Herk would go to Kenia in 1931-1933.

[14]Dutch: Zuiderzeewerken. From the inland sea Zuiderzee large parts were reclaimed from 1920–1970. The first part was the Wieringermeer in the period 1924–1930.

[15]'Congress Programme IAU 1928'; *Hemel en Dampkring*, 26, pp. 208–209, 1928.

[16]The vice-presidents were: G. Abeti (Italy), C. Fabry (France), N. Nörlund (Denmark), F. Nušl (Czechoslovakia), F. Schlesinger (USA).

Fig. 12.1 The executive committee of the general assembly of the IAU in Leiden in 1928. From left to right on the steps of the Observatory: H. Deslandres, F. Schlesinger, president W. de Sitter, general secretary F. Stratton, S. Hirayama, A. Eddington

was his finest hour.[17] Shapley wrote to De Sitter four years later, after the next general assembly, that the dinner in Cambridge had been a bit *dry*. He remembered the *vicious* joys of Leiden well [ASL, 18-10-1932]. He did not mention if the astronomers had prowled around the centre of Leiden having drunk a small glass of jenever too much.

The Bosscha Observatory in Lembang (Dutch East Indies)

Star observations had been made of old from observatories on the Northern Hemisphere. The Cape Photographic Durchmusterung of Gill and Kapteyn was the first extensive undertaking to catalogue on a large scale the stars of the Southern Hemisphere. From the 1820s onwards the English had had the Cape Observatory. After De Sitter's appointment as director of the Leiden Observatory the wish, primarily coming from Hertzsprung, existed to make observations on the Southern Hemisphere too. Two countries were suitable: South Africa and the Dutch East

---

[17]ASL, Obituaries, *NRC*, November 1934.

Indies. The Dutch astronomer Joan Voûte (1879–1963) played an important role in this development. Voûte, born on Java, became, after studies in Delft, observer in Leiden. He did not make it to a doctor's degree. Voûte, not without means, sought a new chance at the Cape Observatory with Hough from 1913, in order to measure double stars. After a couple of years he tried with De Sitter's support to start research at the University of Stellenbosch or at the Union Observatory in Johannesburg, where De Sitter's friend Innes was director. He asked De Sitter to lend him a few small instruments to start the cooperation between South Africa and Leiden in this way. In the end this came to nothing and in 1919 he had the possibility to start working at the Royal Magnetical, Meteorological and Seismological Observatory in Batavia, in his native country. The Dutch meteorologist W. van Bemmelen (1868–1941) was director of this institute. There he received the 7-in. refractor from Leiden. In a discussion between the two rich tea-growers K. Bosscha, whose grandfather had once pleaded in parliament for Kaiser's new Observatory, and his first cousin R. Kerkhoven and Voûte the plan was concocted to build a new observatory. Voûte tried to find the support of the Academy of Sciences. At the time De Sitter stayed in Arosa, and there was a lively correspondence, amongst others between De Sitter, Kapteyn and Hertzsprung, about the best manner to shape the Dutch influence on the new observatory. De Sitter was an advocate for a maximal influence of Leiden, in particular to receive the observational results. Kapteyn supported a shared influence of Groningen and Leiden. With financial support of Bosscha the new observatory was built in Lembang on Java, 15 kilometres from Bandung. In the end Bosscha managed to keep the Dutch influence under control. He realized a big telescope, a double refractor with an objective of 60 centimetres in diameter and a focal length of 10 meters. Voûte became director of the independent Bosscha Observatory. Due to this independence visits by Dutch astronomers to Lembang on a regular basis did not occur. As soon as this became clear, De Sitter chose cooperation with the Union Observatory in Johannesburg as his new target. Voûte stayed director until 1940 and was succeeded by De Sitter's son Aernout (Pyenson 1989; Van der Hucht and Kerkhoven 1982).

Cooperation Between Innes and De Sitter, Johannesburg and Leiden

At the start of 1923 the directors of the observatories in Leiden and Johannesburg, De Sitter and Innes, signed an agreement to cooperate in the mutual exchange of astronomers. The two had met for the first time at the Cape Observatory during De Sitter's learning years with Gill.

Robert Innes (1861–1933), born in Edinburgh and educated in Dublin, became a Fellow of the Royal Astronomical Society at the age of 17. Already in 1878 he wrote a short communication on the satellites of Jupiter to the *Astronomical*

Fig. 12.2 Robert Innes. Drawing from 1920 by the cartoonist Wyndham Robinson

R. T. A. INNES.

Register (Innes 1878), a journal for amateur astronomers. He went to Sydney, Australia, in 1890 and became a wine merchant. But he kept on publishing on astronomy. Double stars were his passion. He wrote to Gill in 1895 asking if there was a vacancy at the Cape Observatory. Gill could only offer him a simple administrative job, which he accepted. There he met De Sitter in 1897. The De Sitter and the Innes families maintained close and very friendly contacts for more than thirty years. In the correspondence from De Sitter to Innes a few rare stamps were often sent to the stamp collector Innes. Innes became director of the observatory in Johannesburg, first a purely meteorological institute, but an astronomical observatory from 1907, the Union Observatory.[18] Prominent fields of research were proper motions of stars, double stars and Jupiter's satellites. Innes was one of the first astronomers who measured the abrupt changes in the rotation of the earth, a result he was very proud of. Although he had been making observations of these changes, the honour of discovery went to the American Brown by a publication in 1926. Innes also discovered the star closest to the earth in 1915: Proxima Centauri.

De Sitter and Innes were frequent correspondents. Innes had a great admiration for De Sitter, he looked up to him for his intellectual capacities, although he was

---

[18]Afrikaans: Unie Sterrewag.

eleven years his senior. For a certain period he tried to get a grasp of the general theory of relativity, but he did not manage to get the hang of the metric tensor $g_{\mu\nu}$. He wished to end the measurements of the Jupiter satellites in 1922, which he had performed during 14 years. Innes went to Europe in 1922, amongst other things to discuss matters concerning the construction of a new 26-inch refracting telescope, which was being built at the factory Grubb Parsons in England. But the discussions went very wrong for Innes and he could not leave London. Not until January 1923 Innes could cross to the mainland to visit his fiend De Sitter. In a confidential letter De Sitter had written to him in October that he would receive an honorary doctor's title [ASL, Letters from Innes, 1919–1923].

Leiden Astronomers to the Union Observatory

At the proposal by De Sitter Leiden University had decided to confer an honorary doctorate to Innes. He was extremely grateful for it, because he had not followed the academic road, but like so many astronomers in the nineteenth century, that of an amateur. De Sitter said in his speech: from musician (Herschel[19]), from clock-maker (Gill), from wine merchant (Innes) to astronomer. The gratitude was mutual. De Sitter was grateful to Innes for two reasons. He had learnt a lot from Innes in South Africa in his two years at the Cape. The two had had innumerable discussions, also about subjects outside astronomy. Moreover, later Innes had done numerous observations of the moons of Jupiter for De Sitter.[20] The honorary doctorate had for De Sitter an additional advantage: it paved the way for a cooperation that was very interesting for Leiden.

One of the articles in the agreement the two observatories signed said that astronomers from Leiden could make observations for longer or shorter periods in Johannesburg, primarily by taking photographs. The other way round South African students could receive part of their education in Leiden. In reality there would be nearly always an astronomer from Leiden at the Union Observatory, and hardly ever the other way round. The Dutch government approved of the agreement already in March, whereas the approval from the Minister of the Interior in Pretoria took some time.[21] Hertzsprung seized with both hands the opportunity the signing of the agreement offered him. At last De Sitter could redeem his promise to Hertzsprung, who saw his long cherished wish come true to study the starry heavens of the Southern Hemisphere. He was the first Leiden astronomer to work in Johannesburg for a long period. He left already in 1923 only to return in 1925. He

---

[19]William Herschel (1738–1822) was a composer and astronomer, who came from Germany to England. He discovered the planet Uranus, the first new planet of modern times.

[20]ASL, Address of De Sitter at the occasion of Innes' honorary doctorate, 24 January 1923.

[21]ASL, Several letters in 1923 between Ministers of both countries, De Sitter and the Board of Governors.

was very happy to have seen the constellation Southern Cross, α Centauri and the Magellanic Clouds in Johannesburg [ASL, 3-12-1923], glad to be able to work without impediments, without a jealous wife, no rheumatism, no cold, no toothache [ASL, 5-2, 6-2-1924]. Hertzsprung and his wife Hetty had decided to divorce in the beginning of 1923. Astronomy was Hertzsprung's only partner in life. Sometimes he even became lyrical:

And then those rich skies with stars, where the pearls are for the taking.

[ASL, 20-2-1924]

He wished to stay longer and the fact that after five years the conversion of his house had hardly been started, served him well as an argument [ASL, 27-2-1924]. Innes wrote to De Sitter that Hertzsprung had a lot of success and that it would be a crime to order him back to Leiden already in May 1924. He hit it off well with everybody and they discovered something new every week. He had to stay at least a whole year [ASL, 27-2-1924]. De Sitter asked permission to the Board of Governors and they agreed. Hertzsprung worked like a Trojan for fifteen hours a day and before his return to Leiden he visited the Second assembly of the IAU in Cambridge in 1925 [ASL, 28-1-1925 from Innes]. But he did not forget the natural beauties either. Amongst other things he visited Transvaal to see the lions and hippopotami, and the waterfalls of the Zambezi river [ASL, 16-7-1924].

His main observations concerned variable stars, of which he tried to ascertain as precisely as possible the form of the intensity curve as a function of time. With these data he tried to find the relation between the period and the form of the intensity curve. He discovered what he tried to find: small differences in the intensity curves correlated with small differences in the period. He sent a few dozen of his measured curves to Eddington, so that he could try to find the theoretical explanation [ASL, 29-4-1925]. This was another of Hertzsprung's important contributions to astronomy. Once, in his enthusiasm, he published a recently discovered variable star twice in the *BAN* [ASL, 21-5-1924]. And he proposed to De Sitter, if he wished to advertise the Leiden results, to say that Hertzsprung had discovered more Algol variables in a single year than anybody else in his or her whole life [ASL, 11-11-1924]. Hertzsprung wrote in his report over the year 1924 to have taken 1267 photographic plates with the Franklin-Adams telescope, with a total exposure time of 446 h, 54 min and 39 s. Amongst others the plates were taken from the Large Magellanic Cloud and the area of star η Carinae. Besides those Hertzsprung had taken over 100 plates to observe small planets and comets [ASL, 7-1-1925].

The second Leiden astronomer at the Union Observatory was Van den Bos, who arrived in Cape Town with wife and child in August 1925. Van den Bos studied in Leiden and worked on his double star measurements under the guidance of Hertzsprung. By Hertzsprung's prolonged stay in Johannesburg De Sitter was his official advisor, when he received his doctor's title in 1925 on his dissertation on *Micrometer measurements of double stars*. After that he went to Johannesburg as Hertzsprung's successor. He stayed for a few long periods in Johannesburg and received an official appointment at the Union Observatory in 1928, as successor of Wood, who in his turn had succeeded Innes as director. It took some difficulties to convince Wood, but Van den Bos' qualities were decisive.

Innes suddenly died in 1933 and De Sitter wrote an obituary in the Monthly Notices of the RAS (De Sitter 1933). De Sitter's other good friend Schlesinger wrote him:

> I enjoyed your writing about Innes. I think I admired him nearly as much as you did, but I also think that I was more aware of his weak points than you were; or perhaps you kept to the adage *nil nisi bonum* (only good things).

[ASL, 9-5-1934]

Later Van den Bos succeeded Wood and was director from 1941 until 1956. His speciality stayed double stars. He had fabulously good eyes to do astronomical observations and measured 74,000 (pairs of) double stars, of which 3000 newly discovered by him. Van den Bos was chairman of the Committee 26 for Double Stars of the IAU for a period of 14 years. Due to illness he had to resign his directorship in 1956 (Muller and Baize 1974).

After the appointment of Van den Bos as chief assistant in 1928 Van Gent was the next Leiden astronomer in Johannesburg. Schlesinger, always in search of talented astronomers, would like to appoint him at the southern station of Yale in South Africa, but in the end that was not possible. De Sitter wrote to Van Gent that if he went to Yale for a certain period, it would not be easy for him to come back to Leiden afterwards; moreover, getting his doctor's title would become a problem [ASL, 12-3-1930]. Van Gent's work on his dissertation developed a bit difficult. Hertzsprung wrote to De Sitter in 1930 from Johannesburg that Van Gent's observational diligence was invaluable, but that he thought too little about becoming a doctor in Leiden [ASL, 12-12-1930]. De Sitter agreed [ASL, 6-1-1931]. Still Van Gent was also active in the theoretical field. He devised a method to determine an ellipse from five observations, without using Kepler's laws. He asked De Sitter's opinion about it [ASL, 17-6-1931]. Van Gent made thousands of photographic plates of variable stars from 1928 until 1946, which were sent to Leiden to be reduced. He was, as Van den Bos wrote, an astronomical miner. Getting rough diamonds from the Transvaal heavens, and leaving the polishing into jewels to others (Van den Bos 1947). Van Gent received his doctor's degree in 1932, with Hertzsprung as advisor, on a dissertation on variable stars. Moreover, he discovered many asteroids, one of which, discovered in 1935, he named after De Sitter. He wrote to De Sitter that he was prepared to hand over the newly discovered small planets to the Union Observatory, but only on the condition of getting something in return. For this commercial thought De Sitter gave him a severe warning. He wrote him it actually was something in return from Leiden as gratitude for the use of the telescope. Van Gent must not lower himself to the commercial level of others, De Sitter let him know. De Sitter would ask Wood to mention Van Gent's name in possible publications [ASL, 4-12-1929]. Van Gent had a certain obstinacy, and De Sitter saw himself obliged to lecture him a second time, referring to a letter from Van Gent to Hertzsprung:

Fig. 12.3 The Observatory staff in 1931, in front of the Observatory entrance

You enjoy a great measure of freedom, but in the end the director or your superior decides.

[ASL, 28-8-1933]

After a successful career as an observer Van Gent died suddenly in 1947 after his return to The Netherlands (Oort 1947). After his death an analysis of observations of five variable stars around β Centauri made by Van Gent appeared, prepared for publication by Oosterhoff.

The Rockefeller Telescope and Leiden Southern Station

In contrast to the development of the larger telescopes, Hertzsprung had the following idea in 1927. At the Union Observatory in Johannesburg, for instance 16 small camera's directed to the pole could be mounted on one single axis (opening of 3 in. and focal length of 20 in.). With that instrument a large part of the skies could be measured every night. Variable stars could then be identified with a blink microscope (or blink comparator). With a blink microscope two photographs of the same area of the sky, taken a few days after one another, could be viewed rapidly interchanging in the same place. Differences between the two plates may then be noticed quickly. For instance: a planetoid is rapidly moving to and fro, or a variable star is rapidly changing in brightness. De Sitter asked Innes if the Leiden instrument

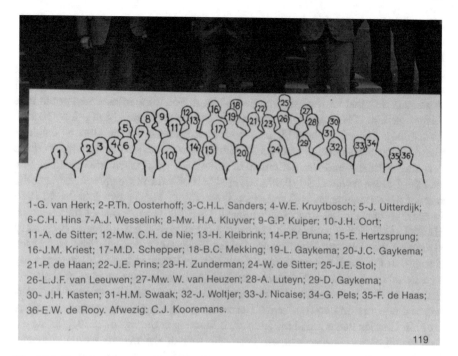

1-G. van Herk; 2-P.Th. Oosterhoff; 3-C.H.L. Sanders; 4-W.E. Kruytbosch; 5-J. Uitterdijk;
6-C.H. Hins 7-A.J. Wesselink; 8-Mw. H.A. Kluyver; 9-G.P. Kuiper; 10-J.H. Oort;
11-A. de Sitter; 12-Mw. C.H. de Nie; 13-H. Kleibrink; 14-P.P. Bruna; 15-E. Hertzsprung;
16-J.M. Kriest; 17-M.D. Schepper; 18-B.C. Mekking; 19-L. Gaykema; 20-J.C. Gaykema;
21-P. de Haan; 22-J.E. Prins; 23-H. Zunderman; 24-W. de Sitter; 25-J.E. Stol;
26-L.J.F. van Leeuwen; 27-Mw. W. van Heuzen; 28-A. Luteyn; 29-D. Gaykema;
30- J.H. Kasten; 31-H.M. Swaak; 32-J. Woltjer; 33-J. Nicaise; 34-G. Pels; 35-F. de Haas;
36-E.W. de Rooy. Afwezig: C.J. Kooremans.

119

Fig. 12.4 Numbers Observatory staff in Fig. 12.3

could be placed on the grounds of the Union Observatory. After that he could elaborate the plan further and try to find funding [ASL, 27-8-1927]. Innes and Wood saw no objections, but warned that a double camera would lead for a number of parts to double costs. In the meantime the number of sixteen had been decreased from sixteen to two [ASL, 21-9-1927]. Two years later the plans for an instrument of Leiden Observatory in South Africa became more real. It had to be a double photographic telescope. Shortly before De Sitter's departure to South Africa in 1929 to visit the meeting of the British Association (BA), he visited the English factory for optical instruments Grubb Parsons in England to discuss the possibilities and prices of the new instrument.[22] In relation to the funding De Sitter had developed the idea to ask a substantial subsidy from the Rockefeller Foundation. He used his trip to South Africa also to gain the support from the South African government for the plans. Money was needed for the building of an observation dome with working rooms for the new Leiden station. At a reception he spoke about it with general Smuts[23], who would become president of the British

[22]ASL, Travel Report South Africa, 1929. (Dutch: Reisverslag Zuid-Afrika.)

[23]General Jan C. Smuts (1870-1950) was Prime Minister of South Africa from 1919 until 1924 and from 1939 until 1948.

Association in 1931. He discussed the matter also with Prime Minister Hertzog[24] during a visit. The lobbying led to the support of the South African government, in the end also financially [ASL, 5-11-1929 from Wood].[25]

De Sitter sent the definitive request for subsidy to the Rockefeller Foundation at the end of 1929. He wrote to Schlesinger that he hoped the decision would be taken in January and that he was very nervous about it. Schlesinger answered not to take further action. That could be too much. It looked good [ASL, 28-12-1929, 14-1-1930]. Schlesinger and Shapley used their influence in the USA as much as possible and reassured the anxious De Sitter, who wrote nervously what to do, a few times. In the end the definitive decision was only taken after another few months. Leiden received $ 110,000.00. First half of it for the telescope and later the rest for the remaining costs. The Leiden Observatory Fund[26] was founded in 1930 in order to manage the Rockefeller money. De Sitter became one of the directors. He had pulled off a major achievement. Leiden would get its own observatory station on the Southern Hemisphere: Leiden Southern Station. A major part of the forming of the ideas had come from Hertzsprung, who had been dreaming already of a southern observatory before his arrival in Leiden. But the process of acquiring the money had been mainly De Sitter's work. He had effectively used his international network on two continents. Hertzsprung and De Sitter discussed the matter with Sir Charles Parsons and his advisors in London at the end of 1930, after which they travelled to Newcastle upon Tyne, the seat of the manufacturing company Grubb Parsons, to draw up a definitive contract for the new telescope.

Then an intensive correspondence with England and South Africa followed. All data and measures of the telescope had to be brought into agreement with those of the new building that would be erected in South Africa. First assistant of the Union Observatory Van den Bos exerted himself enormously to prepare all business in Johannesburg as well as possible. It became a lengthy process. In the meantime De Sitter made his long trip to the USA. At last in July 1933 the building started in Johannesburg. But also the quality of the lenses caused Hertzsprung and De Sitter a lot of concern. Hertzsprung had determined that the lens characteristics of the triplet lenses of the objectives were no better than those consisting of two lenses. In April 1934 Grubb Parsons informed Leiden that the telescope would be ready for inspection in September. De Sitter wrote in July to Wood that he had been in Newcastle. Hertzsprung, De Sitter and his son Aernout would go to England for the definitive inspection and the packing for transport. Aernout would accompany the telescope to South Africa to be present at the erection of the telescope in the new building. And he would officially accept the telescope on account of the Leiden Observatory Fund. Then he would do the first observations with the new telescope. Wood wrote that the building was ready and that he enjoyed the visit of the young De Sitter [ASL, 21-8-1934]. But everything would go differently. De Sitter died at

---

[24]James B.M. Hertzog (1866–1942) was Prime Minister of South Africa from 1924 until 1939.

[25]ASL, Travel Report South Africa.

[26]Dutch: Leids Sterrewachtfonds.

the end of 1934. Moreover, the matter of the lenses had still not been solved. The guaranteed high transparency of the used glass for ultraviolet light could in the end not be realized by Grubb Parsons. Leiden turned to Zeiss in Jena for different lenses. Grubb Parsons agreed with difficulty and it was not until 1938 that the telescope arrived in South Africa, with the lenses of Grub Parsons, which were hired temporarily, because Zeiss could not yet supply them. Hertzsprung and De Sitter made the first observations with the new telescope. Only in 1950 the Zeiss objectives would arrive in South Africa, due to the Second World War and Jena being situated in the German Democratic Republic[27]. In 1954 the Rockefeller telescope was moved to a new plot of ground of the Union Observatory. Due to the disturbing light of the expanding Johannesburg new grounds were chosen in Hartebeespoortdam, north of Johannesburg and 40 kilometres to the west of Pretoria.

Aernout de Sitter would become director of the Bosscha Observatory in Lembang in 1940. After De Sitter's departure to Lembang W. Martin became the Leiden astronomer at the Union Observatory. Later he would go to Lembang too. Both would die during the Second World War as a result of the Japanese occupation.

Equator Expedition 1931–1933

In the years of search to the structure and dynamics of the Milky Way the precise determination of the position of stars was of eminent importance. Distances and proper motions of stars could only be determined on the basis of the most accurate measurements of the positions of stars. It was a thorn in Kapteyn's flesh that of the two co-ordinates of a star, the declination and the right ascension, the last could be measured with extreme accuracy, but the accuracy of the declination measurements was considerably worse. He had signalled the problem already in 1881. Systematic errors of a single observatory were often not more than $0''.1$. But declination measurements performed by different observatories often differed $1''$. The main cause was supposed to be the refraction of light rays by the earth's atmosphere. This refraction is of course dependent on the direction of the incoming light, and of density and temperature variations in the atmosphere.

The amateur astronomer C. Sanders, working in Cabinda, former Portuguese Congo and now an exclave province of Angola, sent articles to *The Observatory* in 1917 and 1919, in which he explained a method devised by himself to use azimuth measurements done near the equator in order to determine the declinations of stars and so evading refraction errors (Sanders 1917, 1919). De Sitter and Kapteyn were positive about the method of Sanders, because extremely accurate determination of star positions was necessary for parallaxes and proper motions, the fundamental

---

[27]Deutsche Demokratische Republik (DDR).

conditions to determine the structure and dynamics of the Milky Way. Kapteyn asked a subsidy of 4500 Dutch guilders from the Holland Society of Sciences (HMW), which was granted. In his request he called the situation concerning the declinations *unbearable*. His aim was to order a new instrument and to use it for measurements at the equator. With supplementary measurements on the Northern and Southern Hemispheres accurate declinations could then be reached. Kapteyn was delighted and thought about the new instrument. He asked De Sitter to take action before leaving to the assembly of the IAU in Rome [ASL, 23-12, 30-12-1921]. But Kapteyn would become ill in 1922. He died on 18 June. So he could not witness the building of the new instrument, and the preparations and results of the expedition to the equator.

Declination determination with Sanders's method
For the definition of the declination see the left figure. The angle between the direction to the star and the plane through the equator is the declination δ.

For the usual measurement of δ we take an observer on the surface of the earth. See the middle figure. In this figure the declination is: $\delta = \beta + \zeta$, with β the latitude of the observer on the earth and ζ the zenith distance of the star. We must realize now that the light from the star bends through the atmosphere in the plane of drawing. However, this bending depends on the place of the observer on earth and temperature differences in the atmospheric layers. This bending is very hard to measure. So the errors are in a high degree unknown. This was the problem Kapteyn had struggled with for years.

The method of Sanders evades the measurement of a declination from an arbitrary place on earth. He chose an observation site on or very near the equator. In the right figure we have an observer on the equator. The direction west-east corresponds with the plane of the equator in the left figure. The telescope of the observer in the right figure is directed in the horizontal plane tangent to the earth and sees the star just above the horizon. The angle

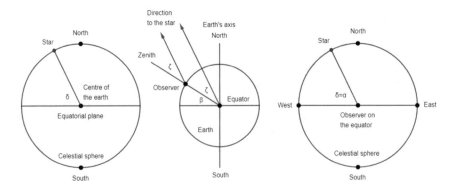

Fig. 12.5 Determination of the declination with Sanders's method

between the direction west and the star is exactly the declination δ, as can be seen in the left figure. But it is also the azimuth of the star in this particular measurement, which is the reason it is called an azimuth instrument. However, there is a fundamental difference with the usual measurement of the declination. The light rays bend through the atmosphere perpendicular to the horizontal plane. But that bending is not relevant. There is no bending which gives an error in the measurement of the declination. In this manner Sanders's method solved the problem of the big errors in the declination measurements.

Altitude, azimuth and zenith distance

The left figure is the plane of the horizon, through the observer and in that point tangent to the surface of the earth. The right figure is the circular section with the celestial sphere of the plane through the observer, the zenith and the star. The zenith is the intersection of the line through the observer, perpendicular to the earth's surface, and the celestial sphere. The projection of the star is the intersection of the circle through the zenith and the star and the horizontal plane.

The angle α the azimuth of the star. We took west for the fixed direction.

The angle λ is the altitude of the star.

The angle ζ is the zenith distance of the star.

Drawn in a single figure the two circles are perpendicular to one another, where the lines observer-projection coincide.

Sanders made a lot of observations with his own instrument, which he sent to Leiden. He had to do everything on his own and actually De Sitter was his only scientific contact. Sanders's education was theoretically weak and it is remarkable

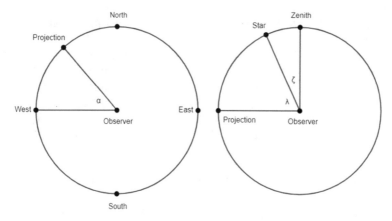

Fig. 12.6 Altitude, azimuth and zenith distance

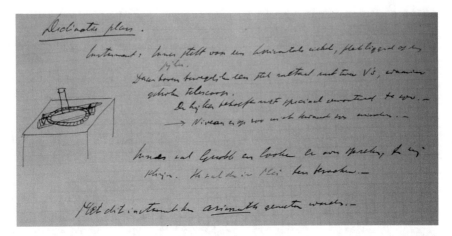

Fig. 12.7 First draft of the azimuth instrument in one of De Sitter's notebooks. Translation: Declination plan. Instrument: Innes proposes horizontal circle, lying flatly on a column. On top a movable piece of metal with two V's in which [lies] telescope. The telescope does not have to be specially mounted. [...] Innes will discuss it with Grubb and Cooke and write to me. I will visit them then in May. With this instrument <u>azimuth</u> can be measured

with how much consideration De Sitter regularly tried to rectify Sanders's misconceptions with elaborate calculations. Sanders's business in Africa went bankrupt and he was in grave financial difficulties. De Sitter paid Sanders for a couple of years roughly 100 Dutch guilders every month as remuneration for his observations. De Sitter let the HMW know in 1922 that the construction of a new azimuth instrument would be too expensive (15,000 Dutch guilders at the Bamberg factory in Berlin). For the time being Sanders would make observations with his own instrument [ASL, 11-10-1922].

De Sitter and Innes, who was in Leiden at the start of 1923 for his honorary doctorate, made their own design for the azimuth instrument. The first sketch is drawn in one of De Sitter's *Study Notebooks*. De Sitter gave the honour of the basic idea to Innes.

Innes, back in England, discussed the design with the Cooke factory and they should send a sketch to De Sitter [ASL, 21-2-1923]. The data, sent by Sanders were analysed by Oort and on that basis De Sitter and Oort formulated a draft scheme of research. Analysis and plan appeared in the BAN (Sanders and Oort 1925) (De Sitter and Oort 1925). In the plan the observation from a place near the equator was essential. After a number of years of observations with his own instrument Sanders came to The Netherlands and was appointed as chief assistant at the Observatory in Leiden in 1926. He was very grateful to De Sitter for this appointment. De Sitter informed the HMW again in 1927. In fact, the research had hardly begun, he wrote, but in the meantime an assignment had been given to Cooke, Troughton and Simms in York (GB) to build an instrument that would cost 8000 Dutch guilders. It would consist of a telescope with an objective of roughly 10 centimetres and a length of

Fig. 12.8 The azimuth instrument

1 m [ASL, 22-11-1927]. De Sitter had written to all his possible sponsors to finance the project. In the end the total funding would be 30,000 Dutch guilders. The two main amounts came from the IAU and the government. The IAU had taken the decision at the general assembly in 1925: 9000 Dutch guilders; the government provided a grant of 5000 Dutch guilders. On top of that the Observatory itself provided 10,000 Dutch guilders, making the total budget 40,000 Dutch guilders. With this money not only the instrument had to be bought, but a long expedition to a spot near the equator had to be paid for as well. The first location considered was Quito in Ecuador in South America, but in the end that location was dropped, because the climate for astronomical observations was not suitable. A place in the Dutch East Indies was also dropped, because of the long-lasting monsoon. In the end Kenia was chosen. The azimuth instrument arrived in Leiden at the end of 1930. The instrument was very heavy, three men were not enough to lift it, De Sitter wrote to Schermerhorn.[28] Schermerhorn, professor of geodesy in Delft, had advised De Sitter in this period. He would check the division of degrees of the instrument in Delft.

In his annual report of 1931 De Sitter called the departure of the Kenia expedition the big happening of the year. The aim was to determine zonal corrections for

[28]Wim Schermerhorn (1894-1977) was professor of geodesy from 1926 and later Prime Minister after the Second World War, from 1945 until 1946.

the declinations. To reach that the declination of 480 stars had to be determined, evenly distributed over the area between −50° and +50°.

Observer Hins, leader of the expedition, and chief assistant Van Herk left by ship from Amsterdam for Mombasa in Kenia, then still a British colony, with over twenty crates of material. Besides the scientific instruments, such as telescopes, mirrors and chronometers, also a radio, batteries, tools, tents and beds, and even two carbines and two pistols were part of the total luggage. As observation post Hins and Van Herk chose the village of Equator northwest of the capital Nairobi, in the middle of nowhere, at 43" south of the equator. They found accommodation at a farm and built a small observation hut, from which the roof could be removed during the nights: The Equator department of Leiden Observatory, as they called it. Ox-carts towed the materials from the railway station at Equator to the farm and the observation hut. The weather was not cooperative for their observations.

Nights with a completely clear sky were rare and there were often rain showers and icy winds. The bad conditions made measurements difficult and influenced the results negatively. In the correspondence between Hins and De Sitter the severe circumstances were not seldom discussed, besides the scientific subjects. Financial matters of Hins were also a recurring subject (what is private and what is business). Another point was the big difference in personality between Hins and Van Herk. The last was a very private young man, who did not speak much and was content with little. Hins was a man in his early forties, temporarily far from his family, who much more needed company and going out. Now and again he travelled to Nairobi to be amongst people for a couple of days and go to the horseraces. It was a lonely existence. Hins suffered from it: working during the nights in not seldom harsh weather conditions, and trying to get some sleep during the days at a farm in full operation. Regularly De Sitter had to hearten Hins, trying to keep his spirits up from afar (Do you play chess often?). After a year and a half of observing De Sitter decided to end the expedition and Hins and Van Herk returned to Leiden in April 1933.[29] The results were not what they had hoped for, but De Sitter was satisfied, because it was a first experiment with the new method. Hins and Van Herk published their catalogue of fundamental declinations in the *Annals of the Observatory in Leiden* in 1938 (Hins and Van Herk 1938). In this article they explicitly praised De Sitter:

> But besides all this, we remember with gratitude professor De Sitter, who not only was the steering mind behind the expedition, but who with his never abating enthusiasm and by his scientific and human advices did hearten us when difficulties occurred and gloominess struck hard.

Hins and Van Herk wrote in their article that they had not solved the problem of the irregularities in the declination measurements rigorously, but that they considered it worthwhile to continue the experiments (Katgert 1991).

---

[29]AUL, Oort archive, Correspondence De Sitter-Hins.

In later years Oort organized a second, more elaborate and longer expedition, which would take place from 1947 until 1951, then with Van Herk as leading astronomer.

References

Blaauw, A. (1994). *History if the IAU, The Birth and First Half-Century of the International Astronomical Union*, Kluwer Academic Publishers, Dordrecht, The Netherlands, 1994.

De Sitter, W. (1933). 'Robert Thorburn Ayton Innes'; *MNRAS*, 94, 4, pp. 277–281.

De Sitter, W. and Oort, J. (1925). 'Provisional scheme for the determination of fundamental declinations from azimuth observations'; *BAN*, III, 81, 1925.

De Sitter-Suermondt, E. (1922). *Report of the trip to Rome 1922*, in the family archive of granddaughter T. van den Eelaart-de Sitter.

Henroteau, F. (1925). 'The International Astronomical Union at Cambridge'; *Journal of the Royal Astronomical Society of Canada*, 19, pp. 197–200, 1925.

Hins, C.H. and Van Herk, G. (1938). 'A Catalogue of Fundamental Declinations Derived from Azimuth Observations at a station on the Earth's Equator'; *Annalen van de Sterrewacht te Leiden* (AOL), XVIII, 1, pp. 1–45, 1938.

Innes, R.T.A. (1878). 'Correspondence – Satellites of Jupiter', *Astronomical Register*, 16, p. 128, 1878.

Katgert-Merkelijn, J.K. (1991). 'The Kenya Expeditions of Leiden Observatory'; *Journal for the History of Astronomy*, 22, 4, pp. 267–296, 1991.

Muller, P. and Baize, P. (1974). 'In Memoriam W,H, van den Bos'; *L'astronomie*, 88e année, pp. 309–311, 1974.

Oort, J. (1947). 'In Memoriam Dr. H. van Gent'; *Hemel en Dampkring*, 45, pp. 159–160, 1947.

Pyenson, L. (1989). *Empire of Reason*. See References of chapter 1.

Sanders, C. (1917). 'Determination of Latitude and Declinations'; *The Observatory*, XL, pp. 271–273, 1917.

Sanders, C. (1919). 'The Determination of Absolute Latitude'; *The Observatory*, XLII, pp. 90–91, 1919.

Sanders, C. and Oort, J. (1925). 'A discussion of the determination of declinations from azimuth measures made near the equator'; *BAN*, II, 76, 1925.

Van den Bos, W. (1947). 'In memoriam dr. H. van Gent'; *Monthly Notices of the Astronomical Society of South Africa*, 6, pp. 34–35, 1947.

Van der Hucht, K.A. and Kerkhoven, C.L.M. (1982). 'De Bosscha-Sterrenwacht, Van thee tot sterrenkunde' ('The Bosscha Observatory, From tea to astronomy'); *Zenit*, pp. 292–300, 1982.

Chapter 13
Last Years

*The assumption of absolute unity in the primordial particle
includes that of infinite divisibility.
The galaxy is but one of the miscalled nebulae revealed to us as
faint hazy spots in the sky.*

Edgar Allen Poe (Poe 1848)

Cosmology Again

The British Association for the Advancement of Science organized its yearly meeting again in South Africa in 1929. De Sitter had been there at the invitation of Gill at a previous meeting in 1905 and he did not let pass the opportunity to visit his beloved South Africa again. After all the management obligations in the Senate of the University and as president of the IAU a bit of rest was very welcome and a scientific travel was exactly what he needed. Before his departure he had corresponded with his English friends in order to travel together. Travelling with these friends afforded the opportunity to discuss scientific problems in all quiet. So De Sitter left at the end of June aboard the Nestor to Cape Town, together with amongst others Dyson, Greaves[1], Eddington, Knox-Shaw[2] and Rutherford[3], for a voyage of nearly three months.[4] They called themselves the Astronomical Society of the Atlantic, like the astronomers, including Kapteyn and De Sitter, had done sailing to Cape Town in 1905. On board De Sitter had long discussions with Eddington and the others, about the arrow of time and irreversibility, about the variable stars of Hertzsprung, about the star streams and the universe of Kapteyn, about that of Shapley and Oort, and also about spiral nebulae. Eddington was still very in favour

[1]William M. H. Greaves (1897–1955) was a British astronomer. He became Royal Astronomer of Scotland and president of the Royal Astronomical Society.
[2]Harold Knox-Shaw (1885–1970) was an English astronomer. He was president of the RAS in 1931–1932. Later he raised funds for an observatory in South Africa and became president of the Astronomical Society of Southern Africa.
[3]Ernest Rutherford (1871–1937) was a New Zealand and British physicist. He worked on the physics of the atom and discovered the later called Rutherford scattering.
[4]ASL, Travel Report South Africa, 1929.

© Springer Nature Switzerland AG 2018
J. Guichelaar, *Willem de Sitter*, Springer Biographies,
https://doi.org/10.1007/978-3-319-98337-0_13

of the De Sitter Universe from 1917, which offered the possibility to explain the nebulae receding from the earth. Dyson advised him not to continue with the satellites of Jupiter. On board they received the message that Rutherford, De Sitter and a few other scientists would receive honorary doctorates from the University of Cape Town. In South Africa De Sitter gave a few lectures, on spiral nebulae, Jupiter's moons and fundamental declinations. He conferred with Innes and Wood about the telescope Leiden wished to erect on the grounds of the Union Observatory in Johannesburg. He also gave a lecture for the Dutch Club.

Probably De Sitter's interest in cosmology, which had been in the background for a decade, was rekindled during the Atlantic voyage by his discussions with Eddington. The increasing number of measurements of galaxies receding from the sun with large velocities made the question of theoretical explanations all the more urgent. The question why De Sitter had not worked on cosmological problems for such a long time is not easy to answer. Eddington's remark that De Sitter had invented a universe and had forgotten about it afterwards is not sufficient (Eddington 1934). His illness, the reorganization of the Observatory and his managerial functions, in short lack of time, give only a partial explanation. More to the point is probably what Oort said about it in 1935, namely that De Sitter was not particularly interested in mathematical speculation, if there was no relation to observations. Only later, when the large receding velocities of many nebulae had been measured, his interest returned (Oort 1935). De Sitter had only the very rough data of three nebulae at his disposal in 1917. There was indeed activity in this field, be it on a modest scale, in the years after 1917. The German astronomer Carl W. Wirtz (1876–1939), who had thoroughly studied De Sitter's work from 1917, wrote a few articles about the relation between distance and velocity of receding nebulae already at the start of the 1920s. In 1922 he wrote about a possible linear relation between luminosity and velocity (Wirtz 1922). The Russian mathematician Friedmann published a solution of the Einstein equations with a time-dependent radius R in 1922, thus introducing for the first time a non-static universe (Friedmann 1922). However, the relevance of this article did not reach the astronomers. The Belgian astronomer Lemaître, who had studied cosmology in England with Eddington, published a critical article on the De Sitter model of the universe in 1925 (Lemaître 1925). Two years later he published an important article, giving a solution of Einstein's equations, describing a universe with an increasing radius, at the same time giving an explanation of the radial velocities of the extra-galactic nebulae (Lemaître 1927). In this article Lemaître introduced the term *expansion de l'espace*, the expanding space or universe.[5] This important article had escaped Eddington and De Sitter a few years. A partial cause was perhaps the fact that it was published in French in the annals of the rather obscure Scientific Society of Brussels. Again two years later, in the beginning of 1929, Hubble published his famous article, in which he announced the linear

---

[5]De Sitter coined the name *Het uitdijend heelal* in Dutch (Hins 1935).

distance-velocity relation of receding nebulae, based on observations with the large Mount Wilson telescope, although he wrote it down with a certain caution (Hubble 1929).

When De Sitter had returned from South Africa he travelled immediately to Paris. There he spoke elaborately with Schlesinger, who had been obliged to cancel his journey to South Africa, about business concerning the IRC and the radial velocities. Concerning the distances of nebulae they discussed the possibility to determine the relative distances from the diameters of the nebulae. De Sitter considered this a better approach than starting from the luminosity. In their ensuing correspondence De Sitter mentioned an article he had written about it in 1924–25, which had not been published. He asked Schlesinger if it was possible to publish a short article about the distances of nebulae for the National Academy of Sciences in Washington. This article, written in April 1930 (De Sitter 1930), appeared in July and made reference to two articles by Hubble, *a great paper*, as De Sitter called it, from 1926 (Hubble 1926) and Hubble's paper from 1929 (Hubble 1929).[6,7] In the 1926 paper Hubble discussed the results of a statistical investigation of the visual magnitudes of 400 extragalactic nebulae. In his paper De Sitter discussed, amongst other things, his idea of measuring the distances based on the apparent diameters of the nebulae, starting with the assumption that these diameters were, like the magnitudes, statistically distributed according to a Gaussian frequency function.

After his travel to the conference of the British Association, and the availability of many new measurements of nebulae, De Sitter was back at the front of the cosmological discussions. He was, like Eddington, thinking about an intermediate solution, between Einstein's Model A and his own Model B.

De Sitter was present at two meetings of the Royal Astronomical Society in London in 1930. In February he spoke about the radial velocities of the extragalactic nebulae and the possibility to explain these with the help of his old cosmological Model B.[8] The report of this meeting made Lemaître take up the pen in order to inform Eddington that he had already given the solution of the problem in his article from 1927. It is surprising that Lemaître's article had escaped Eddington's and De Sitter's attention for three years. It is true, it had appeared in not well-known annals from Brussels, but all the more remarkable is that Lemaître had been present at the general assembly of the IAU in Leiden in 1928. There he had had ample opportunities to hand a copy of the article from 1927 to Eddington and De Sitter. Perhaps the two big men at the conference had no time for the much younger Belgian priest. But after Lemaître's letter in 1930 Eddington was shocked, discovering that he had left unread a copy of Lemaître's article somewhere hidden in his desk. De Sitter received the message from Eddington and wrote to Lemaître on 25 March, that he had read his article with the greatest admiration. He himself

[6]ASL, Travel Report South Africa, 1929.

[7]ASL, Letters from and to Schlesinger from October 1929 until February 1930.

[8]'Proceedings of the Meeting of the Royal Astronomical Society'; *The Observatory*, pp. 37–39, February 1930.

Fig. 13.1 The Belgian
catholic priest and
cosmologist Georges H. J. E.
Lemaître (1894–1966) in
1952. Photograph from the
Lemaître archive

had tried to find a similar solution, but had not found it. In the meeting of the RAS
in June Eddington and De Sitter discussed again about the problem and gave all due
honours to Lemaître: the universe started as an unstable Einstein universe, then
evolved to expand according to Lemaître, in the end developing in the direction of a
De Sitter universe.[9] De Sitter quickly wrote a few articles in 1930, about the
practical astronomical ways to determine the distances of nebulae from their
diameters and their total magnitudes, but also about the theoretical explanations,
starting from Lemaître's work. At the end of July 1930 he wrote to Schlesinger that
he had finished his work on the expanding universe for the time being [ASL,
30-7-1930]. It had held him firmly in its grasp again for a whole year.

But in the years thereafter De Sitter published and lectured regularly about the
expanding universe, also in popular magazines. Also the press often paid attention
to the new phenomenon of the expansion. De Sitter gave a lecture for the combined
meeting of both departments of the Academy of Sciences on 10 April 1933. The
title was: About the expanding universe and the time. He made a real show of it.
Without formulae he described origin, structure and future of the universe. There

---

[9]'Proceedings of the Meeting of the Royal Astronomical Society'; *The Observatory*, LIII, pp. 162–
164, June 1930.

must have been heard a lot of *ohs* and *ahs*. De Sitter tried to provide insight into large distances. With 800,000 pinheads in a row you could reach Leiden's Bush from the Old Church, both in Amsterdam.[10] Another factor of 800,000 yielded the diameter of the sun. And the ratio of a pinhead to the diameter of the sun is in its turn equal to the ratio of the diameters of the sun and the Milky Way. And De Sitter explained the emptiness of our own galaxy as follows. Let a pinhead on Dam Square in the centre of Amsterdam be a star, then the next star is in Rotterdam. And at last: let Amsterdam be our Milky Way, then the city of Utrecht is our neighbour galaxy. After an extremely long time the universe would disintegrate in loose galaxies, which cannot have contact with each other anymore. In a certain sense it would be the end. For the beginning of it all De Sitter chose for an explosion (then not yet named the big bang): everything is clenched together in one point and flees apart with extreme high velocities. The irregular structures of the galaxies and the origin of the solar system could according to De Sitter be explained from the disturbances directly after the explosion.

De Sitter asked himself the following: why do we extrapolate our safe theories valid near to us, so far away in space and time? His answer was: we cannot help doing it. And we take the paradoxical conclusions for granted: The questions are the essence of life for scientists, not the answers. He rounded off with his own philosophy of the wonder of unity in nature:

> The consciousness of immediate contact with the deepest mystery of nature, which is the life-giving principle of science, has grown over the last quarter of a century. The multitude of new discoveries and understandings leads in the end to more coherence and unity in science.

Hubble's Criticism

Hubble wrote a letter to De Sitter on 21 August 1930. Amongst other subjects it dealt with De Sitter's article in the BAN, number 185, about luminosities, diameters, distances and velocities of extragalactic nebulae, which De Sitter had sent him together with a few other articles bearing his name. Starting from the suppositions that galaxies all had roughly the same diameter and total magnitude, their relative diameters and magnitudes were an indication for their relative distances. It was Hubble's opinion that the announced further investigations, following his preliminary article of 1929, were a Mount Wilson matter. And he was angry about De Sitter's noncommittal remark that some astronomers had noted that there seemed to be a linear relation between distances and velocities. Actually he found that De Sitter did not abide by the generally accepted ethics that after a first publication and an announced observation programme further publications were reserved for

---

[10]Dutch: Leidsebosje and Oude Kerk.

Fig. 13.2 Edwin Hubble

himself: Hubble. Objectively Hubble's critical letter was slightly exaggerated. It is true that De Sitter had not mentioned Hubble's article from 1929 in his BAN article (he did mention his 1926 article), but he explicitly mentioned both articles in his article in the Proceedings of the National Academy of Sciences, which had appeared roughly at the same time. Moreover, De Sitter had not only used Mount Wilson measurements, but also observations from a much larger set, often older ones, and he analysed explicitly the relation between diameter and distance, something which Hubble evaded as less reliable. An answer from De Sitter could as yet not be found. In later correspondence the case seemed to have been reduced to reasonable proportions, although Sandage,[11] a student of Hubble, later wrote that Hubble had stayed angry for the rest of his life, because De Sitter had not mentioned him in the article (Sandage 2004).

---

[11]Allan Sandage (1926–2010) was an observer as student assistant to Hubble from 1950 until 1953. He made the first more or less reliable calculations of Hubble's constant and the age of the universe.

At Last America—The Lowell Lectures

De Sitter had nourished the wish to visit the USA for many years, a country with many friends and beautiful telescopes. As a result of his influence a large number of young astronomers had already worked there for longer or shorter periods. A first plan to undertake a long trip to the USA in 1930 could not be realized for various reasons. De Sitter had talked it over with Schlesinger in Paris. The possible subjects on which he could lecture during a lecture tour were already chosen: relativity theory and gravitation, fundamental astronomical constants, Jupiter's satellites. And popular lectures about: time, rotation of the earth, origin of the solar system, size of the universe [ASL, 8-11-1929 to Schlesinger]. In the end the trip was planned for 1931. De Sitter wrote to Schlesinger that as a matter of course he could only come, if his fees would be sufficient to cover the expenses of him and his wife. He asked Schlesinger if there were paid lectures available for him [ASL, 23-2-1931]. Schlesinger asked Shapley for help, who wrote that there was a chance the well-paid Lowell Lectures were available. Schlesinger would not be able to meet De Sitter due to a trip to England and South Africa. For that reason both wished to postpone the trip to 1932. But then the general assembly of the IAU would be in Cambridge (USA) and there would be a run on all well-paid lectures.[12] After all, it would still be the end of 1931, with a full programme starting in the east and ending in Los Angeles.

Before crossing the Atlantic De Sitter and his wife were a few weeks in England. First they attended the extraordinary general assembly of the IAU in London on 19 September, where the statutes of the IAU were changed (Blaauw 1994). They also visited the centenary meeting of the British Association on 26 September and the Faraday Memorial[13] on 30 September. They left from Liverpool on 3 October, in the company of Schlesinger, who had skipped South Africa, aboard the RMS[14] Taconia to the United States.

The program comprised among others visits to Schilt in New York, Schlesinger in New Haven, Shapley in Harvard, Curtis[15] in Detroit and Hubble in Pasadena. The observatories the De Sitters visited were successively in the states of Connecticut, New York, Pennsylvania, Washington, Connecticut, Massachusetts, (Toronto, Canada), Michigan, Wisconsin, Colorado, Texas and California. It was a formidable trip, criss-cross through the United States.

---

[12]ASL, Letters to and from Schlesinger in the months from March until June 1931.

[13]In honour of the physicists Michael Faraday (1791–1867) and James Clerk Maxwell (1831–1879), both famous for their work on electricity and magnetism, two memorial floor stones were placed in Westminster Abbey in London on 30 September 1931, near the grave of Isaac Newton.

[14]Royal Mail Ship.

[15]Heber D. Curtis (1872–1942) was an American astronomer. He had a debate (known as the Great Debate in 1920) with Shapley. The last was of the opinion that the nebulae were small and lay in the outskirts of our own galaxy, while Curtis said they were independent galaxies, large and far away.

From 24 until 26 October they stayed in Washington, where a dinner in honour of De Sitter was given at the Dutch Embassy. And on 26 October, between 12:30 and 13:00 h President Hoover received De Sitter and his wife in the White House, together with the young attaché of the Dutch Embassy J. H. van Royen, later Secretary of State of The Netherlands.[16] They spoke about the small Netherlands and the large America, and about astronomy and America's giant telescopes (De Sitter-Suermondt 1948). It seems that Hoover, when he heard that after their trip the De Sitters would pass through the Panama Canal on their way to visit their son and daughter-in-law, on the spot made a telephone call to Governor Burgess[17] to inform him that a great scientist would pass the Canal.[18]

Thereafter they stayed five days with their good friend Schlesinger in New Haven and then ten days with Shapley in Cambridge near Boston. There De Sitter gave the popular Lowell Lectures,[19] six in all, in which he gave a bird's-eye view of the development of the astronomy. In the first two lectures he treated the history, from Ptolemy and Copernicus to Kepler, Galileo and Newton. The next one was dedicated to William Herschel and the start of the study of the stars. In the fourth lecture De Sitter's master Kapteyn was the central figure, with his local system of stars. The last two lectures were dedicated to the modern developments. Number five about practical astronomy: photography, spectrography and the large telescopes. The last lecture was devoted to the theory of relativity and cosmology. Finally, De Sitter had placed all his knowledge, insights and his own results in a large compilation of two thousand years of astronomy. The lectures were a big success and were covered comprehensively in the press. De Sitter, not being a gifted speaker, fascinated his audience and was applauded each time at the end of a lecture. The six lectures appeared in print under the title Kosmos (De Sitter 1932). De Sitter finished the last chapter expressing his deep admiration for nature and the role of mathematics in it, which had driven him during his whole career, with the following words:

> It is a rather common misapprehension that science, by analysing and dissecting nature, by subjecting it to the rigorous rule of mathematical formulae and numerical expression, would lose the sense of its beauty and sublimity. On the contrary, even the purely aesthetical appreciation, say of a landscape, or of a thunderstorm, is, in my opinion, helped rather than impeded by the knowledge, so far as it goes, that the scientific beholder has of the inner structure and the connection of the phenomena. And the measurement and reduction to numbers, 'pointer readings', as Sir Arthur Eddington says, is not the ultimate aim of science, but its means to an end. By the use of mathematics, that most nearly perfect and most immaterial tool of the human mind, we try to transcend as much as possible the limitations imposed by our finiteness and materiality, and to penetrate ever nearer to the understanding of the mysterious unity of the Kosmos.

---

[16]Archives of the Herbert Hoover Presidential Library and Museum, West Branch, Iowa, USA.

[17]The Panama Canal Zone was in those days still under American authority.

[18]Private communication of De Sitter's grandson L. Ulbo de Sitter, Heusden (province of Noord-Brabant), 2008.

[19]The Lowell Lectures are named after the businessman John Lowell Jr. (1799–1836).

After the Lowell Lectures their travelling went to many observatories, visiting friends and acquaintances. They were tiring weeks, but a few times De Sitter and his wife could recover a couple of days and pass a quiet evening playing bridge with friends.

Pasadena and the Einstein-De Sitter Universe

At the end of December De Sitter and his wife arrived in Pasadena near Los Angeles, where the largest telescope of the world stood on Mount Wilson. Before New Year's Eve Hubble and Einstein arrived. The De Sitters spent the whole month of January in California. De Sitter had the opportunity to see Jupiter through the large telescope on Mount Wilson on 9 January 1932. He had long discussions with Hubble and Humason. M. L. Humason (1891–1972) had recently observed a nebula, using an exposure time of 19 h, which receded from the earth with a velocity of 25,000 km per second. The size of the Milky Way was also a point of discussion. A diameter of 100,000 light-years seemed plausible. There was still much lack of clarity concerning the distances of the nebulae.

De Sitter gave a short series of three lectures in Pasadena, the Hitchcock Lectures[20] in January. While the Lowell Lectures had been historical and intended for the general public, De Sitter lectured in January only on the astronomical aspects of the general theory of relativity and in particular on the expanding universe. It is certain that Einstein was in the audience at the last lecture. De Sitter received a long ovation, it was the last lecture during his long travel. Einstein was full of admiration. *So etwas hab' ich noch nie gehört,* he told his wife. The Einsteins and the De Sitters had adjoining hotel rooms and the two men worked intensively together for a couple of days (De Sitter-Suermondt 1948). An important subject in their discussions was the question how the universe would develop in the future. Would gravitation win in the end and consequently the universe change from expansion to contraction? Or would the expansion go on eternally? This would depend on the material density of the universe. Einstein and De Sitter wrote an article of a single page about this discussion. (Einstein and De Sitter 1932).

First Einstein and De Sitter skipped the cosmological constant, which was not necessary anymore due to the expansion of the universe. Then they wrote that the value and sign of the curvature of the universe could not be deduced from the observational results (the expansion and the density, which were only known with a substantial uncertainty). A positive curvature belonged to a universe which would contract in the end, while a negative curvature caused an infinitely long expanding universe. At a critical value of the mass density $\rho_c = 3H^2/8\pi G$ the curvature would

---

[20]The Hitchcock Lectures have been given since 1909 and were made possible by a bequest of Charles M. Hitchcock.

THE RIDDLE OF THE EXPANDING UNIVERSE EXPLAINED ON A BLACKBOARD
IN CALIFORNIA.
Pasadena , California, January 8, 1932.
Professor Albert Einstein and Dr. Willem de Sitter
find themselves in agreement on the formulae used by
Dr. de Sitter to corroborate his expanding universe
theory in a lecture at the Mt. Wilson Observatory in
Pasadena, California.

Fig. 13.3 Einstein and De Sitter in Pasadena, 8 January 1932. The photograph appeared on the front page of the German magazine Der Welt-Spiegel of 31 January 1932

be zero (H is Hubble's constant, then estimated as 500 km/s per megaparsec, nowadays roughly 10 times smaller; G is Newton's gravitational constant). With zero curvature the universe would just not change to contraction, and stay on expanding at an ever lower rate. In view of the fact that the estimate of the density of the galaxies in the observable part of the universe led to a value of the mass density close to the critical value ρ_c, this universe could be a possibility according to Einstein and De Sitter. It was called The Einstein-De Sitter Universe. Even after more than eighty years there is still no certainty about the ultimate fate of the universe. The increasing speed of the expansion, discovered in the late 1900s, caused by so-called dark energy, together with the supposed existence of dark matter, causing extra gravitational forces, have posed new gigantic problems for astronomers and theoreticians. The great success of his America trip and in particular of the closing weeks in Pasadena was one of the peaks of De Sitter's career. The presence of a number of great colleagues in astronomy in his audience, the discussions about the results of fifteen years of theoretical and practical astronomy, to which he himself had contributed so much, must have been an overwhelming experience for De Sitter, after ten years with hardly any cosmology and the last two years again at the front.

On the way back De Sitter and his wife sailed through the Panama Canal and stayed ten days with son Ulbo and daughter-in-law Ingrid Frick in Maracaibo in Venezuela, where Ulbo worked as a geologist. The trip had been accomplished successfully. The fears the children had felt that the trip would have been too arduous for their father proved to be unfounded. De Sitter was even persuaded to give a lecture for the Caribbean Club. They left for Europe on 2 March.[21]

Hertzsprung Offended

De Sitter had assigned the management of the Observatory to Oort during his absence. Hertzsprung opposed this decision furiously. He felt deeply hurt and called it a repudiation, even a stab in the back. Had the title of deputy director then no meaning at all? The communication between the two men on this matter did not work out smoothly at all. But De Sitter, as usual, kept to his decision. There are many rough copies of letters to the Board of Governors, in which Hertzsprung complained about his treatment by De Sitter. In all these letters he doesn't mention De Sitter by name, but he writes M.D.o.t.O., Mr. Director of the Observatory.[22,23] Actually, it is surprising that De Sitter made this choice, not anticipating Hertzsprung's feelings of humiliation, and when he did become aware of those still persisted in his decision. Although, it was not in De Sitter's character to change a once taken decision. It was only for a short period, just over six months, and during De Sitter's illness and stay in Arosa Hertzsprung had done a good job managing the Observatory, which had hardly been reorganized then. In personal matters Hertzsprung may not have been the most tactical man in his contacts, but his integrity was beyond dispute. Oort was no doubt more well-liked amongst his colleagues and was perhaps, at an age of 34, already more competent than Hertzsprung to lead an organization. But just the fact that so many competent men worked in the well-oiled organization of the Observatory, made a transfer of tasks to Oort and skipping Hertzsprung unnecessary. Perhaps De Sitter made a mistake not realizing the effect of his decision on Hertzsprung, perhaps he even wished to be in a certain way kind to Hertzsprung in taking the administrative tasks out of his hands. Hertzsprung would of course stay director in all scientific matters. And

---

[21]Letter from Ingrid Frick, in the family archive.

[22]Dutch: H.D.v.d.S., Heer Directeur van de Sterrewacht.

[23]AH, C046/4, Archive of Hertzsprung, Aarhus, Denmark.

perhaps he did not want to reconsider his decision at the last moment fearing loss of face. It is guessing at De Sitter's motives. But objectively, De Sitter's decision was not elegant towards someone who had made such an important contribution to the scientific blossoming of the Observatory, and someone to which he owed so much. Oort, who kept De Sitter informed in the USA regularly, was worrying about it and wrote to De Sitter that Hertzsprung had not yet reconciled himself to the situation and would write a letter to De Sitter. After his return De Sitter had to reverse the matter as soon as possible, Oort wrote. He wanted no permanent feud between the two. De Sitter considered it also an unpleasant affair and asked Oort to give his best wishes to Hertzsprung. He hoped that everything would blow over. From the documents it is not clear if the relation between the two was restored after De Sitter's return. Probably not, because at the 25-year jubilee of De Sitter's professorship in 1933, he addressed De Sitter short and businesslike and gave the floor, under the guise of a less command of the Dutch language, quickly to Oort.[24] And after the death of De Sitter in 1934 Hertzsprung did not write an Obituary. The annual report of 1934, written by Hertzsprung, contained only a short retrospective. Hertzsprung wrote about De Sitter's scientific work, that he had found time to do his own important theoretical work, in spite of the amount of administrative work and the concern for the members of the staff. That was all. Hertzsprung was still hurt.

Jubilee and Honours

The Observatory of Leiden celebrated its tercentenary in 1933. But internationally there was no reason for celebration. Hitler had been chosen Chancellor of Germany on 30 January. Although there is little documentation about De Sitter's political ideas, he followed the international political developments closely. He had once written to Dyson in 1916 about Germany and the First World War, which proved his keen interest. The rising fascism during the interbellum period in Germany filled many in The Netherlands, including De Sitter, with concern. The disastrous position of the Jews was already clear then. Their continued existence was in danger. Together with 74 others from non-Jewish circles De Sitter signed a manifesto in 1933 titled the *Disenfranchisement of the Jews in Germany*, which appeared in the daily paper *Het Vaderland* (The Fatherland) on 5 May. The first part of the manifesto read:

> The undersigned, representatives from the most diverse non-Jewish circles of the people of The Netherlands, convinced that the disenfranchisement and oppression of the Jews in Germany will continue undiminished and, according to official statements in the press and on the radio, will not stop before that part of the German people which is of Jewish origin will be deprived of every possibility of continued existence, declare with this manifesto

---

[24]ASL, Correspondence Oort, 1931–1932.

their deeply felt indignation about deeds, which till the present day they had considered not possible, because they mean a frightening and humiliating collapse of a civilization achieved with struggle, and which lead back Europe to the most backward and barbaric times.

The 75 undersigned showed their far-sighted view.

On the occasion of the tercentenary of the Observatory De Sitter had written a short history of the Observatory (De Sitter 1933). The jubilee was celebrated on 6 October with an official meeting with speeches and music. A large company with many international guests had convened in the Large Auditorium, also because De Sitter celebrated the 25th anniversary of his professorship. The Rector of the University D. van Blom opened the row of speakers, followed by De Sitter, who spoke about the history of the Observatory. After De Sitter the Minister of Education, Arts and Sciences H. P. Marchant spoke, who due to the economic crisis had no positive financial message for the University. The ceremonious meeting was concluded by a reception at the Observatory. In the evening there was a more intimate meeting in the Observatory to celebrate De Sitter's 25th anniversary as a professor. He was addressed by Hertzsprung, Oort and the English astronomer Stratton, who had been general secretary of the IAU during De Sitter's presidency. In his speech Oort quoted De Sitter's work with Kapteyn in Groningen, according to De Sitter in those days the headquarters of the army, starting the attack on the fixed stars, then still a very little known field of research. But instead of taking the spoils of that research De Sitter continued in the classical astronomy of gravitation, where new discoveries were least obvious. Oort asked: was the solar system in danger of neglect? Or was the stellar astronomy not complicated and difficult enough? About Jupiter's satellites Oort remarked:

> The charm of these satellites is for me, like for most astronomers, a closed book. Orbits and perturbations are a kind of magic box for you.

He said that many astronomers looked down on the tiny fiddling of the planets, astronomy on the square millimetre, till De Sitter outpaced everybody with the general theory of relativity. By skipping the level of star clusters and galaxies, he examined everything together with his cosmological model of the universe. De Sitter had always given his students much freedom and put his trust in them. Oort also mentioned De Sitter's admiration for the great of the earth, in mind, art and literature, and his enjoying his grandchildren and the flowers in the garden. With admiration Oort mentioned Mrs De Sitter, who made switching from the squabbles of work to the family life so easy.[25] De Sitter's friend Schlesinger had not been able to come and sent a telegram wishing him all the best for the coming 300 years [ASL, 5-10, 24-11-1933]. During the celebration a film was shown about the life at the Observatory and there was music, among others by his son Aernout, who played the cello. From Oort de Sitter received an album with the names of all who offered him a picture to be painted by a portraitist. De Sitter could choose the portraitist

---

[25]ASL, Speech of Oort at the jubilee in 1933. Note of Oort.

himself. In advance Oort had sounded out the painter Jan Sluijters[26] telling him that 1000 Dutch guilders were available. Sluijters would do it with pleasure, but it would cost 1500 guilders, without the frame. Later he lowered the price to 1200 guilders, on the condition that the amount would not be made public. De Sitter chose for Sluijters in November and the portrait was ready in January 1934. The portrait is now in the Observatory in Leiden, which has been housed in the Science Campus to the northwest of Leiden since 1974. In the background a few nebulae have been painted.

The Gold Medal

President Crommelin[27] of the Royal Astronomical Society announced in the meeting of January 1931 that the Gold Medal of the Society had been awarded to De Sitter, for his work on the satellites of Jupiter and his contributions to the theory of relativity. The Gold Medal was awarded yearly since 1824. Many of his colleagues in the field had been awarded the Medal earlier or would receive it later.[28] Therefore De Sitter was invited to give the Darwin Lecture of that year (MNRAS 1931a). Darwin, the son of Charles Darwin, was a renowned astronomer, who had worked on the three-body problem, in particular that of sun-earth-moon. Crommelin delivered the laudation on 13 February, in which he treated the achievements of De Sitter extensively. He presented the Gold Medal to the Dutch ambassador R. de Marees van Swinderen Esq., because De Sitter himself could not be present (MNRAS 1931b). It was a pity that De Sitter could not attend the dinner after the presentation. It was the 836th dinner of the RAS. The fame of the members was roughly proportional to the number of dinners enjoyed. A *centurion* had enjoyed more than 100 dinners. Dyson, the Royal Astronomer, was proud of no less than 230 dinners. A club song was sung and many toasts were proposed to the absent De Sitter. The ambassador thanked and made a pun on Jupiter with *Quod licet Iovi, not licet bovi*. De Sitter's son Aernout, who had been working in England for a few months, was present at the dinner and spoke a speech of gratitude.

De Sitter wrote to Hertzsprung, who at the time was in South Africa, on 6 January that he would be rewarded with the Gold Medal of the RAS. I feel shy and ashamed, he wrote. A main reason would probably be the good reception in 1928 of the general assembly of the IAU [ASL, 6-1-1931]. Schlesinger had received it in 1927. In that year Hertzsprung had heard confidentially that he himself and Sampson had also been nominated [ASL, 21-1-1927]. Sampson received the Medal

---

[26]Johannes C.B. Sluijters (1881–1957) was a Dutch figurative painter, famous for his portraits and female nudes. De Sitter's portrait measures 92 cm × 76 cm.

[27]Andrew C. de la Cherois Crommelin was an English astronomer of French descent. He took part in the 1919 solar eclipse expedition to Sobral and was an expert on comets.

[28]To name a few: Kapteyn (1902), Gill (1908), Eddington (1924), Dyson (1925), Einstein (1926), Schlesinger (1927), Sampson (1928), Hertzsprung (1929), Shapley (1934), Hubble (1940).

in 1928 and Hertzsprung the next year. Two years later it was De Sitter's turn. Again Hertzsprung had heard some information. He stayed at the end of 1930, before leaving to South Africa with Greaves, then chief assistant at the RAS. When Greaves mentioned De Sitter as a possible candidate for the next Gold Medal, Hertzsprung had kept a low profile, but had said that in his opinion De Sitter's work on Jupiter's satellites was of a higher calibre than that of Sampson. Greaves then had let slip the remark, that Samson never should have received the Medal. Hertzsprung advised De Sitter to deliver a good and thorough Darwin Lecture. It would not be easy to show transparencies at De Sitter's subject, but that always brightened up a lecture. They paid well: 3 pence per second. More important still was that Leiden and Mount Wilson were now the only two observatories with two Gold Medals [ASL, 28-1-1931].

More Honours

Besides the Gold Medal De Sitter received many marks of honour. A few have already been mentioned. The main are:

** 1909 Associated member of the Royal Astronomical Society.
** 1925 A royal decoration. The University of Leiden celebrated its 350th anniversary. On this occasion Her Majesty Queen Wilhelmina received an honorary doctorate in law. De Sitter was at the time secretary of the Senate. By royal decree of 28 January 1925 he was appointed Knight in the Order of the Dutch Lion, as was Rector Blok. The president of the Board of Governors, De Gijselaar, was appointed Commander in the Order of Orange-Nassau.[29]
** 1925 Honorary doctorate Cambridge (UK). De Sitter received this doctorate on the occasion of the general assembly of the IAU and his appointment as president of the IAU.
** 1929 The Watson Medal. At the end of 1928 De Sitter was awarded the 9th Watson Medal by the American National Academy of Sciences in Washington.[30] The medal is named after the American astronomer James C. Watson (1838–1880). This medal is awarded every three to five years to a prominent astronomer. Gill and Kapteyn were two of his predecessors. Schlesinger asked De Sitter to come and receive the medal in person, but De Sitter wished to visit the USA later [ASL, 27-11-1928].
** 1929 Foreign member of the American National Academy of Sciences. Schlesinger congratulated him and wrote that there were only fifty foreign members [ASL, 6-5-1929].

---

[29]Newspaper article in the *Leidsche Courant*, 10 February 1925.

[30]*Leidsche Courant*, 24 December 1928.

** 1929 Honorary doctorate from the University of Cape Town. De Sitter heard of it during the voyage to South Africa to the BA meeting. He received the doctorate on 1 August.[31]

** 1931 The Bruce Medal. De Sitter received the 26th gold Catherine Wolfe Bruce[32] Medal from the Astronomical Society of the Pacific during a meeting of the Board of Governors of the University of Leiden on Saturday 13 June. This medal existed since 1898 and is awarded at the recommendation of the directors of the six largest observatories in the world. Gill and Kapteyn had also received the Bruce Medal.[33]

** 1931 Honorary doctorate of the Wesleyan University in Middletown, Connecticut, USA.[34]

** 1931 Honorary member of the American Astronomical Society.[35]

** 1932 Honorary doctorate of Oxford University.[36]

** 1933 Honorary doctorate of the Brown University in Providence, Rhode Island, USA [ASL, 9-3-1933].

** 1933 Foreign correspondent of the French Academy.

** An impact crater on the moon is called De Sitter.

** 1935 Posthumously a planetoid, discovered by Van Gent, was named after De Sitter.

De Sitter Dies in November 1934

Still on 17 October 1934 De Sitter wrote to the Academy in connection with a lecture he would give about the relativity theory and the expanding universe [NAW1, 17-10-1934].

His son Ulbo and daughter-in-law Ingrid stayed temporarily at the Observatory. Ulbo had been declared unfit to work abroad and they were looking for a home. Ingrid wrote to her parents in Sweden, that they had the real Spanish influenza[37] in the house. First Ulbo, then De Sitter, followed by granddaughter Ragni. Eleonora, De Sitter's wife, and Ingrid hurried from one sick room to another. Ulbo recovered, but little Ragni died on 16 November of pneumonia. Daughter Agnes travelled from Normandy to Leiden and wrote to her husband about these days. They did not dare

---

[31]ASL, Travel Report South Africa, 1929.

[32]Catherine Wolfe Bruce (1816–1900) was an American philanthropist, particularly in the field of astronomy.

[33]*Leidsche Courant*, 15 June 1931.

[34]Newspaper article in *Het Vaderland*, 16 November 1931.

[35]*Het Vaderland*, 16 November 1931.

[36]*Het Vaderland*, 23 June 1932.

[37]The real Spanish influenza was a pandemic illness in the years 1918–1919, when 20–50 million people died.

tell the dangerously ill De Sitter that his granddaughter had died. When he asked himself if she had died, Eleonora answered *Yes*.[38] De Sitter also died of pneumonia, a few days after his granddaughter, on 20 November. Willem de Sitter and his granddaughter were cremated in the crematorium Westerveld in Velsen (province of North-Holland). Their last resting place is in the urnfield next to the crematorium. The funeral service was held on Saturday 24 November. Many from the scientific world came to pay their last respects. The Rector of the University W. van der Woude, Oort and Huizinga spoke.[39] His old friend Huizinga said:

> To have been one of the many he had admitted into the affection of his great and simple heart, stays, in remembrance for each of us allowed to say it, a treasure for the rest of our lives.

De Sitter was commemorated nationally and internationally in the astronomical world. The chairman of the faculty spoke of a masculine mind in a fragile body.[40] Oort spoke about the miraculous equilibrium in De Sitter's life and about his mysterious manner to face nature. In the eyes of Oort the De Sitter family was the happiest he knew. Moreover, De Sitter was extraordinarily brave, as was always the case with people who did not care about themselves.[41]

Not long after De Sitter's death his widow moved house. She stayed in Leiden and took some plants and flowers with her from the Observatory garden. She wanted to make her new garden again one of the most beautiful in Leiden, with violets, her favourite flowers. She visited the flower market every week. But she had also her pronounced opinions and in later years she meddled in the upbringing of her grandchildren, for instance about the books they could or should not read. Sometimes that led to small frictions with her children(-in-law). She was also socially active, with her own tea parties, lecturing at the Society of Public Advancement,[42] and her weekly hour reading to children from the neighbourhood.[43] Her book about her husband was one big homage to her deceased husband. The first lines were:

> In the beginning. Further and further it recedes, when I gaze to the beginning. When did I first hear the rustling wing-stroke of his mind? Imperceptibly the light has come near, which changed the face of the world.

(De Sitter-Suermondt 1948)

It is remarkable that eighty six years earlier a certain A. Wispelweij wrote a poem in honour of Kaiser, the builder of the Observatory, starting with:

---

[38]Letters from daughter-in-law Ingrid Frick, in the family archive.

[39]*Het Vaderland*, 25 November 1934.

[40]AUL, Faculty of mathematics and physics.

[41]ASL, Obituaries, note of Oort.

[42]Dutch: Maatschappij tot Nut van't Algemeen.

[43]Private communication of De Sitter's grandson L. Ulbo de Sitter, Heusden (province of North-Brabant), 2008.

Fig. 13.4 The gravestone in the urnfield in Westerveld: of De Sitter, his granddaughter Ragni, his daughter-in-law Ingrid Frick and his wife Eleonora. Photograph taken by the Author

He is a great man, who with wings of the mind, may rise to the boundless universe of Heaven.[44]

The widow asked Oort to write a few pages about *De Sitter in his study* [ASL]. After some thought he did not comply with the request, saying in a diplomatic manner that it would not be appropriate to add something to such a personal document. Eleonora de Sitter died on 20 November 1952, exactly 18 years after her husband.

The eldest daughter Theodora (born in 1900) studied French, went to work as a teacher in South Africa and married there in 1924. The couple Smit-de Sitter later emigrated to the USA.

The eldest son Lamoraal Ulbo (born in 1902) studied geology, received the doctor's title in 1925 and married the Swedish Ingrid Frick. For several periods he worked abroad in the field of the extraction of oil. Two of those countries were The Dutch East Indies and Venezuela. When the family was in The Netherlands in

---

[44]Archive of MEOB, Marine Electronic and Optical Company (Dutch: Marine Elektronisch en Optisch Bedrijf); Dutch Institute for Military History (Dutch: Nederlands Instituut voor Militaire Historie), The Hague.

between jobs of Ulbo, they stayed in the big Observatory home of Ulbo's parents. Their little son Ulbo (born in 1930) was always allowed to fetch his grandfather for lunch. Therefore he had to pass the *holy door* to the Observatory and softly knock on the door of his grandfather's study. De Sitter loved it to be guided by his grandson to lunch.[45] Lamoraal Ulbo was appointed at Leiden University in 1935, where he later became a professor. After Ingrid's death in 1939 he remarried Catharina Maria Koomans.

The other son Aernout (born in 1905) studied astronomy. He received a gold medal of the University of Leiden for a prize essay about RZ Cassiopeiae. His father was proud of him and informed *The Observatory*, which mentioned it in a short note.[46] Aernout was interested in philosophy and music, he played the violoncello well, he read Spinoza's *Ethics* with his good friend Oort, he observed birds in the dunes and was a seeker of the deeper meaning of things (Oort 1946). He was definitely a son of his father. He married Hette Zoetelief Tromp in 1930. His relation to his advisor Hertzsprung reached a low in 1935. In a letter to his friend Oort he unloaded his wrath. Suddenly Aernout had to perform another series of observations, which in his own view were unnecessary for his doctorate and would certainly take much extra time [AUL, Oort archive]. He was furious with Hertzsprung and felt humiliated. After his doctorate in 1936 he was the Leiden astronomer at the Union Observatory in Johannesburg from 1937 until 1939, where he made observations of variable stars. He succeeded Voûte as director of the Bosscha Observatory in Lembang on Java. After the Second World War in the Far East had started, he was interned by the Japanese oppressors. In the internment camp he was a support for his fellow prisoners. In a labour camp near Palembang on Sumatra he died in 1944. His colleague-astronomer from Leiden, W. C. Martin, who had been appointed in Lembang in 1941, also died there in 1945. A third Leiden astronomer working in the Dutch East Indies, J. Uitterdijk, who had gone to the Dutch East Indies to become a teacher on Java in 1939, probably drowned when the transport ship he was on, sailing from Batavia (Djakarta) to Malaysia was torpedoed in 1944 (Oort 1946). After De Sitter's death an article appeared in the *BAN*: A study of the variable stars in Messier[47] 4, which had been prepared by Oosterhoff. The article was based on more than a hundred photographic plates of the spherical star cluster M4, which had been sent to Leiden after the Second World War (De Sitter 1947).

Daughter Agnes (born in 1907) married a brother of Aernout's wife. The couple Zoetelief Tromp-De Sitter moved to France, where they had a chicken farm. De Sitter and his wife stayed with their daughter and son-in-law in France a few times.

---

[45]Private communication of De Sitter's grandson L. Ulbo de Sitter, Heusden (province of North-Brabant), 2008.

[46]*The Observatory*, LVI, p. 320, 1933.

[47]The French astronomer Charles Messier (1730–1817) published a list of more than 100 far objects (nebulae, star clusters), numbered M1, M2, etc.

De Sitter's Succession

After De Sitter's sudden death the succession became an awkward question (Katgert-Merkelijn 1997). In University circles Oort, who already had made his name with his articles on the structure of the Milky Way, was tipped by all as De Sitter's successor. De Sitter would retire at the age of 70 in 1942 and Hertzsprung would not be eligible then at his age of 69. But in 1934 Hertzsprung considered himself the only proper candidate for the combined function of professor and director of the Observatory. Once he had tried to create a two-man directorate, but De Sitter had prevented it.

Within the faculty, in the Observatory, as well as in the Board of Governors there were supporters of the directorship of Oort. It was not Hertzsprung's age, he was 61 then, but his lack of social skills which led many to a choice in favour of Oort. Many feared that director Hertzsprung would create much dissatisfaction and animosity. Hertzsprung was an excellent astronomer with a great record of service, but he could make people very angry. Oort himself had many doubts about taking his chance already, which may be inferred from the many small notes in his fine writing, in which he put down the pros and cons of his own directorship versus that of Hertzsprung. The deadlock in which the succession threatened to land, led again to the consideration of splitting the Leiden astronomy: Hertzsprung ordinary professor and Oort director of the Observatory and extraordinary professor. Objectively this was no consistent policy: from H. G. Van de Sande Bakhuyzen alone, via De Sitter and E. F. van de Sande Bakhuyzen shared, De Sitter alone, now a division again? The fact that Hertzsprung could hardly be passed over induced the president of the Board of Governors first to speak with Hertzsprung and Oort and then ask for an audience with Minister Marchant of Education, who then summoned Oort and Hertzsprung. The Minister was slightly in favour of splitting the astronomy in Leiden again, but Hertzsprung persevered: a professor in physics had of course to be also director of the laboratory where his research was done. The Minister was convinced and decided to appoint Hertzsprung professor and director of the Observatory, and Oort extraordinary professor. Oort was allowed to move into the director's house. De Sitter's widow wrote a letter of congratulation to Hertzsprung [AH, C046/3, 31-5-1935].

In the end, the Second World War brought the activities on the Observatory partly to a standstill. Hertzsprung could continue working undisturbed due to his Danish nationality, but Oort had to go into hiding a couple of years on the Veluwe (a nature reserve in the east of The Netherlands). After the war Oort came back quickly and in fact took on the direction of the Observatory. Hertzsprung was nearly 72 and retired. Oort became director, as De Sitter had wished it.

A small group of students, friends and colleagues bade farewell to Hertzsprung on 25 June 1946. Oort addressed him and compared his capacities of observation with those of his Danish predecessor Tycho Brahe (1546–1601). There was no astronomer for whom he had so much admiration as for Hertzsprung, Oort said. Herzsprung had always made his own preferences secondary to the interests of astronomy. Beforehand, an astronomer could never know which measurement was important and which was not. In Hertzsprung's own words (Strand[48] 1947):

> If you work hard, you always find something, and sometimes something good.

Hertzsprung went back to Denmark to live in the village of Tolløse, 40 km west of Copenhagen. The inhabitants were surprised at the peculiar man working for hours and hours in front of the window on an incomprehensible machine. Hertzsprung kept on working and keeping in contact with colleagues, also young astronomers. He visited the general assemblies of the IAU, also the one in Berkeley (California, USA) in 1961, at the age of 87. Martin Schwarzschild (1912–1997), whose father had invited Hertzsprung to come to Göttingen in 1908, gave a lecture about star evolution. He addressed Hertzsprung and mentioned the great theoretical importance of the Hertzsprung-Russell diagram published half a century earlier. Hertzsprung was cheered loudly. He died in 1967 and left his body to medical science (Herrmann 1994).

Concluding Remarks

Although outside astronomy there was little time left for De Sitter, he tried together with his wife Eleonora to pay attention to art and literature in those spare moments. They had the habit of reading aloud to each other from philosophical works. And if there was some time left, they visited museums or the theatre. They stayed in a hotel in The Hague around Christmas in 1932, where they visited exhibitions and the theatre. They went an evening to their son Ulbo and daughter-in-law Ingrid, who lived in Wassenaar then, near The Hague, to play bridge. It was the time that bridge was conquering the western world. The couple felt a great love for nature. In South Africa, when they had just met, they enjoyed the beautiful scenery. As a boy, De Sitter had enjoyed the woods and moors around Arnhem. He also loved riding, skating and cycling. Later De Sitter and his wife, when they were staying in the parental home, had loved going for long walks around Arnhem. It was not only simply enjoying the scenic beauty, the couple felt one with nature in a particular manner. In De Sitter's lectures his deep feelings about the unity of nature came up regularly. In many reminiscences of colleagues the harmony of their family life was mentioned.

In his heyday De Sitter appealed to the imagination of the general public. Even the English satirical magazine *Punch* mentioned him on 12 September 1923:

---

[48]Kaj A. G. Strand (1907–2000) was a Danish astronomer. He worked as an assistant to Hertzsprung in Leiden from 1933 until 1938. Then he went to the USA and later became director of the US Naval Observatory.

Professor W. de Sitter of the University of Leyden, has propounded a theory that the earth is not a rigid body, but a *wobbling jelly*. We can only conclude that he has never seen Bayswater.[49]

De Sitter's Awful Handwriting

De Sitter's handwriting was sometimes very hard to decipher. It was as if he was always in a hurry. Even in letters to the Minister or the Board of Governors of the University he did not try to write well-cared-for letters. Once he wrote a letter to Pannekoek in *Bussum* (in the province of North-Holland). This letter arrived in Bussum with a long delay, due to the unclearly written place-name. In had been sent to *Bremen* in Germany first [ASL, 12-9-1921 from Pannekoek].

We can say about De Sitter, with a reference to Newton, that he had stood on the shoulders of giants, to see so far into the universe. In his lifetime there were four giants whose shoulders he had used. First there is Kapteyn, whose first and most prominent student he was. He was the first to partly unravel the structure of the universe (then only the Milky Way) and the greatest astronomer of The Netherlands around 1900. From Kapteyn De Sitter learned the indivisibility of theory and observation. He was also a student of Gill, one of the most accomplished British practical astronomers. From Gill De Sitter learned the instruments, the technique of observing and how to reduce the observations. In a later period there were Lorentz and Einstein, with whose help he could develop his cosmological ideas, in direct discussions or by letters. He was a lucky scientist with such excellent masters and colleagues.

We can establish a difference between De Sitter's methods of investigation in his two main fields of research: the new analytical theory of the four Galilean moons of Jupiter and the cosmological consequences of the general theory of relativity, i.e. the De Sitter Universe.

De Sitter's joy in mathematics led him to the choice for celestial mechanics, and to the development of a new analytical theory, based on the previous work of great theoreticians from the 19th century, culminating in the work of Poincaré. But he did this work alone, without the help from or discussions with anybody. He was watched with awe by his fellow astronomers, from whom many did not dare try to understand it. The absolute will to find his way in the darkest caves of the analytical celestial mechanics kept him enchanted for thirty years. Many colleagues did not understand the joy of this solitary journey. It was best worded by Oort, calling the fascination for the four moons of Jupiter a closed book for many astronomers.

---

[49]Perhaps the wobbling marshy moors of Bayswater were meant.

In his cosmological work De Sitter steered a different course. Cooperation and discussion with other scientists formed the basis: with Einstein, Lorentz, Ehrenfest, Droste, Fokker and a number of others with whom he corresponded.

Then there is another element playing an important role throughout De Sitter's career, the wish to test every theoretical result at the hand of observations. Because, only if the observations corresponded with the predictions of the theory, it was really worthwhile. The large series of observations of the satellites of Jupiter, performed at many observatories at the request of De Sitter, were reduced by him and his students, resulting in ever better values of the masses and orbital elements of the satellites, and in ever decreasing unexplained residues. For Jupiter and its moons De Sitter stood in a tradition of three centuries of studying this system, in fact started with the discovery by Galileo of the four largest moons. De Sitter was one of the last great theoreticians in analysing the complex mutual influencing of the large and small celestial bodies of our solar system.

De Sitter also tried to find experimental confirmation in his cosmological work, first in 1917 when only a few data were available. With only three galaxies of which the velocities were reasonably known, he tried to ascertain the effect he had found of the receding velocities of test masses in his cosmological model. De Sitter's waning interest in cosmology during the 1920s may perhaps be partly explained by the lack of sufficient data about the distances of the nebulae.

Summarizing we may argue that De Sitter's rare combination of talents as a theoretician as well as a practical astronomer, together with his wish to put the theory always to the test of the measurement results, made him a great scientist. There is a clear unity in the work of the three great Dutch astronomers: Kapteyn, De Sitter and Oort, covering nearly a century: the unravelling of the structure of the Milky Way, the complex of the galaxies and finally the structure of the universe.

Amsterdam, 1 January 2009 (Dutch version), 24 March 2018 (partly translated and extended version in English).

Jan Guichelaar.

References

Blaauw, A. (1994). *History of the IAU*. See References of chapter 12.

De Sitter, W. (1930). 'On the Distances and Radial Velocities of Extra-Galactic Nebulae, and the Explanation of the Latter by the Relativity Theory of Inertia'; *Proceedings of the National Academy of Sciences of the United States of America*, 16, 7, pp. 474–488, 1930.

De Sitter, W. (1932). *Kosmos*, Harvard University Press, Cambridge, Massachusetts, 1932.

De Sitter, W. (1933). *Short History of the Observatory of the University at Leiden, 1633–1933*, Joh. Enschedé en Zonen, Haarlem, 1933.

De Sitter, A. (1947), 'A study of the variable stars in Messier 4 (Partly discussed and prepared for publication by P.Th. Oosterhoff)'; *BAN*, X, 385, pp. 287–303, 1947.

De Sitter Suermondt, E. (1948). *Een menschenleven*. See References of chapter 3.

Eddington, A.S. (1934). 'Obituary'. See References of chapter 11.

Einstein, A. and De Sitter, W. (1932). 'On the Relation between the Expansion and the Mean Density of the Univers'; *Proceedings of the National Academy of Sciences*, 18, pp. 213–214, 1932.

Friedmann, A. (1922). 'Über die Krümmung des Raumes'; *Zeitschrift für Physik*, 10, pp. 377–386, 1922.

Herrmann, D.B. (1994). *Ejnar Hertzsprung*. See References of chapter 10.

Hins, C.H. (1935). 'In Memoriam W. de Sitter'; *Hemel en Dampkring*, 33, pp. 3–18, 1935.

Hubble, E. (1926). 'Extragalactic nebulae'; *Astrophysical Journal*, 64, pp. 321–369; *Mount Wilson Contributions*, 324, 1926.

Hubble, E. (1929). 'A Relation Between Distance and Radial Velocity Among Extra-Galactic Nebulae'; *Proceedings of the National Academy of Sciences*, 15, pp. 168–173, 1929.

Katgert-Merkelijn, J.K. (1997). 'De opvolging van W. de Sitter' ('The succession of W. de Sitter'); *Jaarboekje voor geschiedenis en oudheidkunde van Leiden en omstreken (Annual of the history of Leiden)*, pp. 128/143, 1997.

Lemaître, G. (1925). 'Note on De Sitter's Universe'; *Journal of Mathematics and Physics*, IV, 1, pp. 188–192, 1925.

Lemaître, G. (1927). 'Un Univers Homogène de Masse Constante et de Rayon Croissant, Rendant Compte de la Vitesse Radiale des Nébuleuses Extra-Galactiques'; *Annales de la Société Scientifique de Bruxelles*, XLVII, pp. 49–59, 1927.

(MNRAS 1931a) *Monthly Notices of the Royal Astronomical Society*, XCI, 3, p. 258, 1931.

(MNRAS 1931b) *Monthly Notices of the Royal Astronomical Society*, XCI, 4, pp. 422–434, 1931.

Oort, J. (1935). 'In memoriam'. See References of chapter 6.

Oort, J. (1946). 'Obituary, W.Chr. Martin, A. de Sitter, J. Uitterdijk'; *The Observatory*, 66, pp. 265–266, 1946.

Poe, E.A. (1848). *Eureka*; Hesperus Press Limited, London, 2002 (First published in 1848).

Sandage, A. (2004). *Centennial History of the Carnegie Institution of Washington, Volume I, The Mount Wilson Observatory*, Cambridge University Press, New York, 2004.

Strand, K.A. (1947). 'Ejnar Hertzsprung and the Leiden Observatory'; *Popular Astronomy*, 55, p. 361, 1947 (containing a translation of Oort's speech).

Wirtz, C. (1922). 'Einiges zur Statistik der Radialbewegungen von Spiralnebeln und Kugelstarnhaufen'; *AN*, 215, pp. 349–354, 1922.

Index

© Springer Nature Switzerland AG 2018
J. Guichelaar, *Willem de Sitter*, Springer Biographies,
https://doi.org/10.1007/978-3-319-98337-0

Printed in the United States
By Bookmasters